向慕尼黑学习

— 对历史城区可持续发展的思考 —

Learning from Munich : Rethinking Sustainable Development of Historic Districts

刘　崇　李舍悦　著

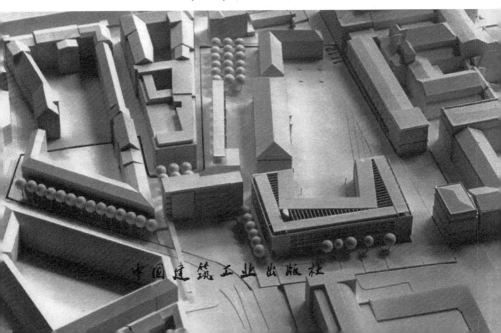

中国建筑工业出版社

图书在版编目（CIP）数据

向慕尼黑学习：对历史城区可持续发展的思考 =
Learning from Munich: Rethinking Sustainable
Development of Historic Districts：英汉对照 / 刘
崇，李含悦著 . —北京：中国建筑工业出版社，
2022.12（2024.4 重印）
ISBN 978-7-112-28175-6

Ⅰ.①向… Ⅱ.①刘… ②李… Ⅲ.①城市规划—城
市管理—研究—慕尼黑—英、汉 Ⅳ.① TU984.518

中国版本图书馆CIP数据核字（2022）第227449号

责任编辑：何　楠　徐　冉　段　宁
责任校对：党　蕾

向慕尼黑学习
对历史城区可持续发展的思考
Learning from Munich: Rethinking Sustainable Development
of Historic Districts

刘　崇　李含悦　著

＊

中国建筑工业出版社出版、发行（北京海淀三里河路9号）
各地新华书店、建筑书店经销
北京海视强森文化传媒有限公司制版
北京中科印刷有限公司印刷

＊

开本：880 毫米 ×1230 毫米　1/32　印张：4　字数：126 千字
2023 年 4 月第一版　　2024 年 4 月第二次印刷
定价：**39.00** 元
ISBN 978-7-112-28175-6
　　（39939）

在本书中，作者选择以巴伐利亚州首府慕尼黑为例，展示什么是好的城市建设。这一选择基于两个重要的假设：第一，慕尼黑非常适合于展示城市设计和城市管理的"最佳实践"。这个假设得到了国际城市排名的支持，慕尼黑在这些排名中常常位列第一。第二个假设涉及跨文化的维度。作者假设研究欧洲的城市设计和管理有望获得更多的知识。他们认为，对慕尼黑的关注有助于启发和提升中国的城市设计。显然，这种跨文化假设包含这样一种观念，即存在着一些典型的城市设计方法、程序、技术和目标，它们可以从德国的文化环境中汲取，在中国的文化环境中发挥作用。

"智慧城市发展的常量与变量"论点特别令人着迷，它提出将城市跨文化的以及本土文化的"常量"和"变量"分离开来进行研究。作为这个理论的例证，正如本书表述的那样，刘崇不仅在德国学习并获得魏玛包豪斯大学的博士学位，而且在时间允许的情况下，他总能在德国建筑师和规划师事务所发挥作用。他曾作为著名建筑师贡特·海茵公司的员工来到慕尼黑，有很多时间来了解这座城市及其重要的建设项目。

让我们举一个例子来说明城市发展的"常量"与"变量"之间的联系。对于好的城市住区，在任何文化环境下都要具备的"常量"包括公共交通和私人交通的有效运作、方便抵达的商业服务设施与休闲场所，气候友好型的建筑设计和城市规划等。与此相对应的是，慕尼黑历史特色浓郁的周边式街坊，虽然不严格强调住宅的朝向，但通过其界面对公共空间的围合强调了城市的文化价值。

这向中国读者传达了一个关于城市规划的信息：要实践良好的规划，不必复制、模仿或移植慕尼黑的城市文化，而是从慕尼黑的城市建设中独立于其文化的优秀案例中获得灵感就足够了。这些灵感可以在中国的实践中通过适当的文化转译而发挥作用。

让我们更详细地探究一下，什么是好的城市建设的标志？在我看来，作者的研究表述了五个特点。

第一个标志，是审慎保存与延续城市的历史、建筑记忆以及以空间所能表达的城市叙事。这些叙事总是与当地的传统深深地交织在一起。例如在欧洲，它们与围合公共空间的周边式街坊，或与市政厅和教堂作为"城市之冠"的文化传统交织在一起。城市可能遭受过破坏，但好的城市建设可以治愈、维持和延续城市的空间构成。只有这样，空间的记忆才能继续存在，城市的品质才能传承给未来。二战后德国城市的重建中，对待传统的不同态度导致了不同的结果。下萨克森州首府汉诺威以汽车优先、强化功能分区的现代规划为重建方向，而慕尼黑则选择了相对更灵活、现代与传统共生的方式来重新塑造城市。作者更加认同慕尼黑的发展模式，因为它不仅重视城市功能和社会品质的提升，也强调城市形象的美学价值。

第二个标志涉及城市空间的社会构成。好的城市建设始终关注城市空间的社会生活品质。开放的或公共的城市空间，首先应该能够超越其使用功能，服务于人们的互动，促进人们的共享。这些空间还应充分反映人口的结构，为各种生活方式的人们提供发展机会。而且，它还应在不限制多样性的前提下帮助缩小社会阶层的差距。

第三个标志，如作者所讲，是合理的组织和安排城市空间，持续地促进经济发展和地方繁荣。这首先包括高效的基础设施，也包括促进创新的教育机构或有吸引力的文化场所。如果城市交通效率低下，或缺乏有活力的商业和居住空间，靠减税就能促进繁荣吗？

好的城市建设的第四个标志，是特别关注建成环境对城市生态的影响。例如，被硬化土地对气候的影响、通风廊道效能的发挥、污水与垃圾处理设施的能力、流动或静止水体的休闲功能，以及以公园和花园等形式存在的"城市绿肺"。好的城市建设不仅把城市看作一个社会系统，而且始终也把城市看作一个生态系统。例如，促进生物多样性与加强城市应对灾难的韧性都发挥着重要的作用。

最后，也是第五个标志，形象的建设不能缺失——这个形象会一再出现在当地居民以及国内外访客的脑海中。在德国，"笔记本电脑和皮裤"是巴伐利亚州及其首府慕尼黑倾心打造的形象设定，其隐含的意义是"在这个地区，现代与传统合二为一"。

在某种程度上，有吸引力的城市形象整合了前文提到的好的城市建设具备的所有特征，并将它们组合成一个连贯的信息。这个信息作为"潜台词"，让人联想到城市的标志性建筑或景观（如慕尼黑的圣母教堂、奥林匹克体育场、玛利亚广场等），提供着关于居民归属感、身份认同感、职业前景、生活质量，以及城市经济和社会繁荣程度等方面的资讯。好的城市形象，必然是高水准的城市建设文化的反映。

　　我高兴地接受了刘崇的邀请，为本书作序。其中的重要原因是我相信，这里记录的案例研究将吸引学术青年和热情读者，以及建筑和城市设计领域的实践专家。他们将从中领会到优秀的城市建设在文化、历史、经济、社会和生态等领域的多元化目标。

迪特·哈森福鲁格

2022 年 10 月 9 日，魏玛

The study you hold in your hands centers on the topic of successful urban design. For this purpose, the authors choose the Bavarian capital city of Munich as prime example. His choice is linked to two significant hypotheses. The first is based on the assumption that Munich is, in fact, well suited to demonstrate "best practice" in the realm of urban design and city management. This assumption is supported by the results published in international city rankings that regularly feature Munich in top echelons.

The second hypothesis relates to the intercultural dimensions of the topic. The authors suggest that studies of urban design practice and city management oriented on European cities hold the promise of gaining valuable knowledge – also for Chinese actors. Deliberating on the case of Munich, as the author states, has the potential to inspire and hence, to advance urban design. Quite evidently, such intercultural assumptions also support the recognition that a toolbox typical to the profession exists. This toolbox comprises methods, procedures, techniques and aims that allow their detachment from one cultural milieu (Germany) and subsequent attachment to another cultural context (China).

From this vantage point, theories on invariants in smart urban development gain specific importance. They imply a strict separation between invariant, unchangeable and variant, changeable dimensions of the urban way of life, dimensions that extend beyond cultures and those that are particular to local cultures. In this regard and as the study demonstrates, Chong Liu stands on firm ground. For one, he studied in Germany, where he received his doctoral degree (at the Bauhaus University Weimar). In addition, time and again, when the opportunity arose and his schedule permitted it, he worked in German architectural and urban design firms. One such opportunity led to joining the team of renowned architect Prof. Dr.-Ing. Gunter Henn at his Munich office, where the author spent a significant amount of time and became acquainted with the city and its most important projects.

Let us take a look at an example in order to illustrate the interrelation between invariants and variants in urban design. For instance, the intercultural requirements that successful urban housing development needs to meet include the provision of efficient public and private modes of transport, proximity to shopping facilities and services (short travel distances), access to nearby recreational areas, climate-friendly equipment of buildings or spaces and many more. In contrast, a cultural characteristic of the city of Munich such as block border construction is quite oblivious to cardinal directions, yet emphasizes how public urban space is articulated and, thus, clearly constitutes a local cultural practice.

As a result, urban design formulates a message that is addressed to its Chinese readers: In order to practice successful urban design, it is neither necessary to copy, nor to imitate, nor to import the spatial culture of Munich. Feeling inspired by the interculturally transferable, exemplary practices of urban design in Munich is completely sufficient. Such inspirations permit their recontextualization in Chinese urban design and, in follow—after experiencing a process of cultural adaptation and modification—they can unfold their effectiveness.

We should therefore ask: What actually characterizes successful urban design? In my point of view, the authors study reveals five key characteristics.

The first one concerns the respectful preservation and gentle modernization of urban history, its built memory or, in other words, the spatially defining narratives of cities. These narratives are deeply intertwined with respective local traditions. In Europe, for example, this refers to how the culture of block border construction dramatizes public space or the historic "City Crown" comprising market place, city hall and church. Time and again, cities are subject to disruptions of their history while being capable of tolerating such disruptions. Successful urban design has the capacity to "heal" the spatial composition, to foster it, and to continue its story. This is the only way how spatialized memory can live on, and how urban qualities of place can be handed down through time. The appreciation for this tradition is demonstrated by the different concepts that drove the post-war reconstruction of German cities following the end of World War II. For instance, the rebuilding of Hanover, the capital city of the state of Lower Saxony, was based on the guiding image of the carfriendly, functionally differentiated modern city. The city of Munich, on the other hand, chose an undogmatic and flexible type of historic reconstruction of the urban image. Chong Liu points out that he prefers the Munich approach to rebuilding, for functional and social reasons, but also due to the aesthetic advantages this approach affords to the image of the city.

The second characteristic relates to the social and societal constitution of urban space. Successful urban design will always pay attention to the qualities of spaces in cities in relation to the urban way of life. In addition to their functional performance, open or public urban spaces should enable, even promote human interaction. They should further represent the demographic composition of the population in an appropriate way by offering opportunities for personal development, and for self fulfillment. Finally, they should contribute to the mitigation of social disparities without limitation of societal diversity.

Third, according to the authors, successful urban design is qualified by its capacity to organize and dramatize urban space in a manner that allows fostering economic development and local prosperity in a sustainable way. Most of all, this includes efficient infrastructure and spaces required for an educational system capable of promoting innovation, as well as an attractive range of cultural activities. Of what use are low business taxes if systems of mobility or accessibility are deficient or attractive areas for commercial and housing development are insufficiently designated?

Fourth, successful urban design can certainly pay precise attention to the ecological implications of the built transformation of the environment. This relates, for instance, to the climate impacts of coverage, the interruption of fresh air corridors through buildings, the performance of sewage infrastructure and waste processing, the natural conditions of standing or flowing water suitable for local recreational purposes, and last but not least, the "green lungs" of cities in the form of parks and public gardens. Successful urban design considers cities not only as a social system, but also as an ecosystem. Enabling biodiversity plays an important role, just as strengthening the resilience of urban spaces in anticipation of disasters of any kind.

Finally, fifth, the image of the city must not be forgotten–an image that appears in the minds of individual local urbanites, as well as those of visitors from abroad. An image that is commonly known in Germany and refers to both Bavaria and Munich posits a union of computers and traditional leather shorts: "Laptop und Lederhose" has, thus, become a metaphor for the marriage between modernity ("Laptop") and tradition ("Lederhose").

In a certain way, an attractive urban image integrates all previously described characteristics of successful urban design and orchestrates them within a coherent message. The text of this message, ingrained as subtext within the architectural beacons of the city (Frauenkirche church, Olympic Stadium, Marienplatz central square, etc.), tells us stories of belonging, identity, social harmony, professional outlook, prosperity and quality of life. It also heralds the high quality of local urban building culture.

When Chong Liu asked me to write this preface, I was happy to comply. One reason is that I am convinced that the case studies documented here will attract an enthusiastic readership among academic youth, as well as experts in practice, in the fields of architecture and urban design. At the same time, they will gain an awareness of the complex cultural, historic, economic, social and ecological goals of a local urban building culture of exemplary quality.

Dieter Hassenpflug
Weimar, 9th October 2022

目录
Contents

序 | Preface

引言 | Foreword 001

1. 保障经济繁荣 | Supporting economic prosperity 007
让创意生态灿若繁星 | Attracting creative companies 010
提供高效的政府服务 | Improving governmental efficiency 014
保持居民和企业容量 | Keeping residents and enterprises 016

2. 彰显城市文化 | Promoting urban culture 018
让大学繁荣开放共享 | Letting universities open and shared 022
新旧共生的建筑修复 | Renovating with old and new elements 024
寓旧于新的城市体验 | Providing rich urban experience 026
先进的科研建筑设计 | Designing advanced buildings 028
划破天际的歌剧空间 | Offering awesome music space 030
与时俱进的住区规划 | Updated residential design 032
包容丰富的建筑色彩 | Tolerating architectural colors 038
成熟社区加建养老院 | Adding elderly care in mature communities 040
废弃用地变示范小区 | Revitalizing abandoned area 043
塑造场所的城市设计 | Building active spaces 047
创造交流和聚会机会 | Creating togetherness 051
让科技展览享誉世界 | Exhibiting world-level collections 053
构筑公共文化的地标 | Promoting cultural highlight 055
黄金地段建人才公寓 | Establishing talent housing in golden location 059

3. 倡导社会整合 | Advocating social integration 061
使衰落住区重生活力 | Activating declining areas 064
兼顾公平的土地开发 | Developing with social equality 066
让公众参与规划决策 | Decision-making with participation 068

4. 保护自然生态　|　Protecting ecological environment　069

　建设大规模生态公园　|　Developing large parks　072

　生态修复硬化的河道　|　Naturalizing concrete river beds　074

　为骑行人士提供便利　|　Giving bikers convenience　076

5. 示范绿色科技　|　Demonstrating energy efficiency　078

　公示建筑能效证书　|　Implementing energy certificate　081

　太阳能小区储能示范　|　Storing solar energy in housing　083

　博物馆的精细化调控　|　Adjusting museum indoor climate　086

　教堂建筑的采光设计　|　Design natural light in church　090

　老王宫的棱镜形屋顶　|　Introducing daylights in parlarment　094

　实现通风防噪的外墙　|　Ventilating while avoiding noise　096

　更新近现代文物建筑　|　Regenerating modern heritage buildings　098

6. 保障建筑品质　|　Ensuring building quality　100

　优秀的教育培训体系　|　Teaching and training with good system　103

　专业开放的项目论证　|　Debating projects in open platform　105

　利益相关方良性互动　|　Interacting healthly among stakeholders　107

参考文献和图片来源　|　Literature and picture source　109

后记　|　Postscript　112

研究历史城区的可持续发展，需要兼顾功能、美学与社会维度的发展质量，也需要有国际视野的本土创新。英国都市生活杂志 *Monocle* 的"2018年全球最宜居城市排行榜 Top25"上，慕尼黑荣登榜首。根据全球人力资源机构 Mercer 公司 2019 年的研究，慕尼黑不仅是世界上生活质量最高的城市之一，也是德国最适宜居住的城市。在多维度上取得高质量的可持续发展，正是本书选取慕尼黑为研究对象的原因。

慕尼黑作为地名在 1158 年的文献中首次被提及，原意为"修士之所"，后从聚居点演变成为建有防御性城墙的巴伐利亚公国首府。17 世纪后城市步入快速发展的阶段，19 世纪以前的慕尼黑已是建有内外两道城墙和巴洛克式宫廷建筑群的区域行政中心。19 世纪的慕尼黑拆除外城墙，建设城市环路，不断融合周边的小城市和村镇，直至 1967 年市域范围不再扩张。今天的慕尼黑（Stadt München）人口逾 150 万，划分为 25 个市辖区（Stadtbezirke）。

Studying the sustainable development of historic districts requires a balance between functional, aesthetic and social aspects of development quality, as well as local innovation with an international perspective. Munich topped the list of the British urban life magazine *Monocle*'s "The Most Liveable Cities in The World". According to a 2019 survey by Mercer, Munich is not only one of the cities with the highest quality of life in the world, but also the most livable city in Germany. The high-quality development in multiple dimensions is precisely the reason why this book chose Munich as the object of study.

Munich was first mentioned as a place name in 1158, originally meaning "the place of the monks". It evolved from a settlement to the capital of the Duchy of Bavaria with defensive walls. After the 17th century, the city entered a period of rapid development, and before the 19th century Munich was a regional administrative center with two inner and outer walls and a Baroque court complex. In the 19th century, Munich demolished the outer city wall to build an urban ring, constantly integrating the surrounding small cities and villages until 1967. Today the city of Stadt München has a population of more than 1.5 million and is divided into 25 municipal districts.

慕尼黑鸟瞰

Stadtbezirke München

1. Altstadt-Lehel
2. Ludwigsvorstadt-Isarvorstadt
3. Maxvorstadt
4. Schwabing-West
5. Au-Haidhausen
6. Sendling
7. Westpark
8. Schwanthalerhöhe
9. Neuhausen-Nymphenburg
10. Moosach
11. Milbertshofen-Am Hart
12. Schwabing-Freimann
13. Bogenhausen
14. Berg am Laim
15. Trudering-Riem
16. Ramersdorf-Perlach
17. Obergiesig
18. Untergiesing-Harlaching
19. Thalkirchen-Obersendling-
    Forstenried-Fürstenried-Solln
20. Hadern
21. Pasing-Obermenzing
22. Aubing-Lochhausen-Langwied
23. Allach-Untermenzing
24. Feldmoching-Hasenbergl
25. Laim

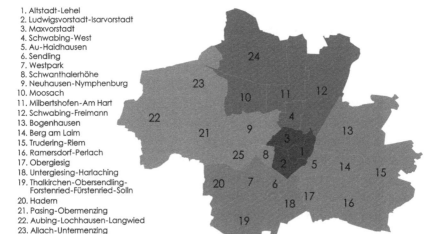

慕尼黑历史城区范围示意图

这里保留着众多的传统建筑与文化遗存，洋溢着往昔巴伐利亚王国首府的历史风情，又被亲切地称为"百万人的村庄"。慕尼黑也是本书中"慕尼黑历史城区"所指的范围。它是慕尼黑规划区（Planungsregion München）的中心，也是包含奥格斯堡、因戈尔施塔特和罗森海姆等周边城市的慕尼黑大都市区（Metropolregion München）的中心，是德国南部最现代和最繁华的大都市，是仅次于柏林和汉堡的德国第三大城市。慕尼黑优雅的古都风范和丰富的历史建筑给人们留下深刻而美好的印象，其实今天的慕尼黑是经历了两次大的城市重建的结果。第一次城市重建是 15 世纪中后期，经历了大规模火灾和鼠疫破坏后；第二次重建则是因为慕尼黑曾是纳粹德国的政治中心和经济重镇，二战中受到盟军空袭的重创，城市中心区九成建筑被毁，许多美轮美奂的古建筑变成一片瓦砾。战后市政府当局面临两种选择，即采用当时国际上流行的现代主义风格进行快速重建，或恢复建筑与街道、广场原有的传统风貌。我们今天看到的慕尼黑就是第二种选择的结果。

The City of Munich is also the scope of the "Munich Historic District" in this book. Preserving many traditional architectural and cultural relics, full of the historical style of the former capital of the Bavarian Kingdom, it is affectionately known as the "village of a million people". The City of Munich is the center of the Planing Region of Munich and the heart of the Munich Metropolitan Region, which includes the surrounding cities of Augsburg, Ingolstadt and Rosenheim. It is the most modern and prosperous metropolis in southern Germany and the third largest city in Germany after Berlin and Hamburg. Munich's elegant historical atmosphere makes a deep and beautiful impression, but today's Munich is the result of two major urban reconstructions. The first urban reconstruction was in the mid-to-late 15th century, after large-scale fire and plague destruction. The second reconstruction was due to the fact that Munich was once the political center and economic center of Nazi Germany, and was hit hard by Allied air raids in World War II. Ninety percent of the buildings in the city center were destroyed, and many beautiful ancient buildings were reduced to rubble. After the war, the municipality was faced with two options, namely to quickly rebuild in the modern style that was popular at that time, and to restore the original traditional style of the buildings, streets and squares. The Munich we see today is the result of the second option.

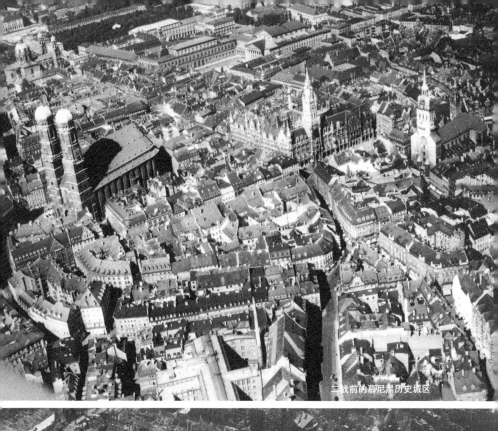
二战前的慕尼黑历史城区

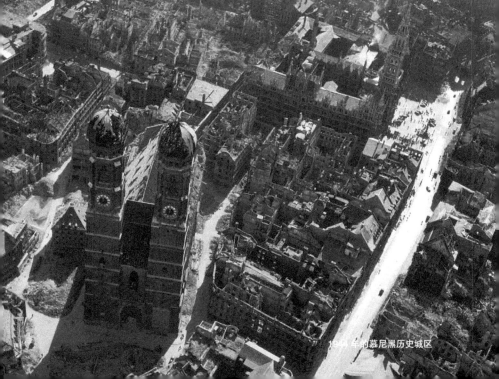
1944 年的慕尼黑历史城区

慕尼黑历史城区的修复并不是完全意义上的"修旧如初"。在"古老"的建筑立面的背后，基础设施和内部建筑空间的设计完全面向着现代化发展的需求，而新的建筑形式也常常与历史建筑并置。这种兼顾古都风貌和未来发展的重建模式取得了令人瞩目的成功，被很多学者称为"慕尼黑道路"。本书呈现的六个原则，即保障经济繁荣、彰显城市文化、倡导社会整合、保护自然生态、示范绿色科技以及保障建筑品质等，正是笔者从设计师和学者视角对"慕尼黑道路"的解读。

本书面向对建筑设计、城市设计、历史城区保护和更新以及中西方比较研究领域感兴趣的读者。慕尼黑历史城区位于北纬48°，日照条件与我国的哈尔滨相似，各月份平均气温又接近山东、河北地区的城乡，因此它在规划与建筑层面的发展对我国诸多北方城市非常具有借鉴价值。

The restoration of Munich is not exactly "restore the old building". Behind the "old" façade, the infrastructure and interior architectural spaces are designed to be completely oriented to the needs of modern development, and new architectural forms are often juxtaposed with historic buildings. This reconstruction model, which takes into account the style of the ancient capital and its future development, has achieved remarkable success and has been dubbed the "Munich Way" by many scholars. The six principles presented in this book, namely prospering the economy, highlighting culture, promoting fairness, protecting the ecology, coexisting with the old and new, demonstrating new technologies, and ensuring the quality of construction, are the author's interpretation of the "Munich Way" from the perspective of a scholar.

This book is intended for readers interested in the fields of architectural design, urban design, preservation and renewal of historic districts, and comparative studies of China and the West. The City of Munich is located at 48° north latitude. Its sunshine conditions are similar to that of Harbin in China, and the average temperature of each month is close to that of Shandong, Hebei and Henan provinces. In this sense, its urban regeneration has much reference value for many northern cities in China.

1. 保障经济繁荣
Supporting economic prosperity

慕尼黑充满既古朴又变革的力量，这里的设计和创意产业蓬勃发展，不少企业的影响力辐射欧洲乃至全球。如果说汉堡聚集着德国企业中的"老贵族"，慕尼黑则更受德国"创意人群"的青睐。慕尼黑的城市管理者清晰地认识到，历史城区拥有吸引知识生产与交流的"创新性氛围"，而营造这种氛围需要好的大学、宜人的环境和多元的社会生活。慕尼黑有 14 家大学和应用技术学院等高等教育机构，近 9 万名大学生，其中慕尼黑工业大学、慕尼黑大学常常和瑞士的苏黎世联邦理工大学共同位列世界排名最领先的德语区大学。历史城区中不仅有各学院的研究所，还有马克斯·普朗克研究院（简称马普研究院）、弗劳恩霍夫协会和国家环境与健康研究中心等著名科研机构，以及多所紧密结合工业界的技术培训学校。此外，慕尼黑还提供了一系列优秀的国际学校、良好的城市基础设施和各种各样的娱乐活动——这一点对年轻的外籍人士特别有吸引力。

Munich is both quaint and transformative, with a thriving design and creative industry and a number of companies radiating to Europe and the world. If Hamburg is home to the "old aristocracy" of German companies, Munich is more popular with the German "creative crowd". Munich's city managers are well aware that in order to retain scientific institutions and talents, it is crucial for the Historic District to have an "innovative atmosphere" that attracts knowledge production and exchange. Creating this atmosphere requires good higher education, a pleasant environment and a pluralistic social life. Munich has 14 universities, universities of applied sciences and other higher education institutions, with nearly 90,000 students, of which the Technical University of Munich, the University of Munich are among the world's top German-speaking universities. There are not only the research institutes of various universities, but also the Marx-Planck Institute, the Fraunhofer-Gesellschaft and the State Environmental and Health Research Center, as well as a number of technical training schools closely integrated with industry. In addition, Munich offers an excellent range of international schools, good urban infrastructures and all kinds of recreational activities—which are particularly attractive for young expats.

没有围墙的慕尼黑大学

慕尼黑工业大学图书馆

城市的发展离不开自然科学与社会科学技术的进步，慕尼黑历史城区经济与社会的繁荣也离不开知识的赋能。慕尼黑因其开放包容的城市性格与多样的创业机遇，是年轻人向往的实习和定居之所。这里既有像宝马、西门子和英飞凌这样的大型跨国公司，也有众多的由三五个人组成的设计工作室。在春夏季的周末，各学院、机构、企业和博物馆还举办"公众开放日"活动，主动邀请市民、学生前来参观和交流。开放日活动不仅全程有专人免费讲解，还常常有精心准备的咖啡和茶点。在这里，专家、市民和学生们在轻松的气氛中交流知识与见解，求职者也可以了解到不同单位的专业度与工作环境。笔者曾工作过的慕尼黑海茵建筑事务所是一个有代表性的大型建筑设计企业，员工近 300人，是德国工业建筑和教育设计领域的领头羊。在研究生产、办公和教育建筑的功能组织与科学设计方面，海茵建筑事务所创立了一整套名为"设计策划"（Programming）的工具体系，得到德国企业界的高度认可，也受到包括西湖大学等中国委托方的青睐。

The development of the city is inseparable from the progress of natural and social science and technology, and the economic and social prosperity of the historic district of Munich is also inseparable from the empowerment of knowledge. Because of its open and inclusive urban character and diverse entrepreneurial opportunities, Munich is a place for young people to study and settle down. There are large multinational companies like BMW, Siemens and Infineon, as well as numerous design studios of three or five people. On the weekends of spring and summer, colleges, institutions, enterprises and museums also hold "public open day" to invite citizens and students to visit and have an exchange. "Public open day" are not only fully narrated by a dedicated person, but also often have carefully prepared coffee and refreshments. Here experts, citizens and students exchange knowledge and insights in a relaxed atmosphere, and job seekers can also learn about the professionalism and working environment of different units. Henn Architects in Munich, where I worked, is a representative large-scale architectural design company with nearly 300 employees, and is a leader in the field of industrial architecture and educational design in Germany. In the study of the functional organization and scientific design of production, office and educational buildings, Henn has created a complete set of tools called "Programming", which is highly recognized by the German business community and Chinese clients like West Lake University, etc.

新绘画博物馆的公众开放日

海茵建筑事务所中标的西湖大学

提供高效的政府服务
Improving governmental efficiency

　　随着互联网与移动通信技术的迅速发展，在线的城市公共服务发挥着越来越大的作用。在德国诸多的城市在线服务平台中，muenchen 网是用户访问量最大的平台之一。它是慕尼黑市政府、上巴伐利亚行政区政府与多个行业协会和银行共同成立的合资运营平台，2004 年正式上线。在 muenchen 网上，可以找到从景点信息到公交路线的全部城市服务的公开信息与在线沟通入口。为了适应外国人与儿童、老年人等不同群体使用，muenchen 网还不断进行着优化与升级，也在社区与学校中宣传和培训使用方法。这一全方位公众服务平台的设立，使得很多企业与个人的经营申请和信息报送可以在线和无纸化地进行，从而起到节省社会公共资源、促进环境保护的作用。从城市形象而言，政府通过可视化手段宣传慕尼黑宜居、宜业的生活品质，公示与城市发展相关的信息，传达本地区发展的愿景与目标，展现了主动运用新科技处理公共事务的能力，提升了城市的亲和力。

With the rapid development of the Internet and mobile communication technology, online urban public services are playing an increasing role. Among the many urban online service platforms in Germany, website of muenchen is one of the most visited platforms by users. It is a joint venture platform established by the Munich Municipal Government, the Government of Upper Bavaria and a number of industry associations and banks, and was officially launched in 2004. On this website, public information and online communication portals for all city services, from tourism information to bus routes, can be found. In order to adapt to the use of different groups such as foreigners, children and the elderly, the website of muenchen is constantly optimizing and upgrading, and is also promoting and training the use methods in communities and schools. The establishment of this all-round public service platform enables many enterprises and individuals to apply for business and submit information online and paperlessly, thus saving social public resources and promoting environmental protection. In terms of the image of the city, the government demonstrates the ability to actively use new technologies to deal with public affairs, which enhanced the affinity of the city.

muenchen 网实时更新公交车次信息与站点建设项目信息

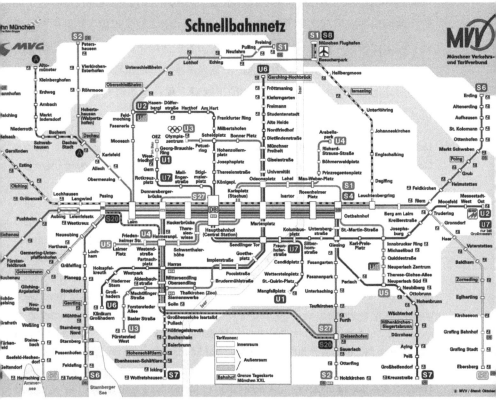

muenchen 网提供的城市交通与票价图

保持居民和企业容量
Keeping residents and enterprises

　　因独特的历史风貌和繁华的商业氛围，慕尼黑历史城区是商务办公与零售业的上乘选址，而政府也一直着力把这里营造成有稳定居住人口的生活区。为此，慕尼黑政府的做法是保持其既有的多样性与包容性——包括保证产业业态与文化设施的种类与持久性，并维持公共机构与私人企业的数量。近 20 年来，历史城区及周边建设了五宫廷（Fünf Höfe）、老庭院 (Alter Hof)、宝马博物馆（BMW-Museum）以及每年有 200 万访客的加斯泰格文化中心（Gasteig Cultural Center）等新的商业、文化地标，还完成了后文将提及的农夫湾与特蕾莎高地等现代化的居住区，并持续改造位于历史城区的公共交通设施。为了促进少数族裔的发展与文化上的包容，特别重建了犹太教堂和犹太社区中心，新建了犹太博物馆。慕尼黑历史城区保持居民的生活场所的属性，可以使街道空间和社会生活拥有持久的活力。通过一系列举措，慕尼黑历史城区的居住人口在经历了 20 世纪 70 年代至 90 年代的下滑后，现在基本维持稳定，而历史城区从未失去其作为城市文化高地和优选居住环境的地位。

Due to its unique historical flair and bustling business atmosphere, Munich's Historic Centre is a prime location for business offices and retail, and the government has been working hard to make it a living area with a stable population. The approach is to maintain its existing diversity and inclusiveness – including the variety and durability of industrial and cultural facilities, and to maintain the number of public institutions and private enterprises. In the past two decades, new commercial and cultural landmarks such as the Fünf Höfe, the Alter Hof, the BMW-Museum and the Gasteig Cultural Center, which receives 2 million visitors a year, have been built in and around the historic district, as well as modern residential areas, and the public transportation facilities have been continuously renovated. In order to promote the development and cultural inclusion of minorities, the synagogue and the Munich Jewish Centre were rebuilt in particular, and a new Jewish Museum was built. The historic district maintains the properties of the living places of the residents and can make the street space and social life have lasting vitality. Through a series of initiatives, the population of Munich's historic district has remained largely stable after a decline from the 1970s to 1990s, and the historic district has never lost its status as a cultural highland and preferred living environment in the city.

充满生活气息的历史城区

2. 彰显城市文化
Promoting urban culture

慕尼黑是拥有丰富历史遗存的古城，而 1942 年至 1945 年遭受到的 70 余次轰炸让城市几乎被夷为平地。二战后多数市民主张以历史的面貌重建旧城、恢复原有的生活品质。为应对城市未来发展的需要，慕尼黑并不拘泥于对历史上城市的原貌进行"整体复原"，而是不断提升建筑功能、优化城市结构、塑造广场和步行街等公共空间的品质、提升建筑的舒适性与使用功能。比如在修缮市政厅的同时，有意将其东西两侧的建筑在重建时后移了 4~5 米，以强化市政厅与整个圣母广场的艺术表现力。政府大规模地恢复了传统的街道系统、主要历史建筑的立面造型与细部，对王宫的恢复一直持续到 20 世纪 70 年代。通过采取一系列经济与民生并举的城市建设举措，慕尼黑得以迅速地重现繁荣的景象、重振市民的自信，并成功地获得 1972 年的奥运会举办权。以奥运会为契机，慕尼黑开始大规模地建设地铁系统，并逐步将老城墙内的整个区域开辟为步行区。前者进一步推动了城市的现代化水平与经济的活力，经改造而闻名遐迩的步行区也成为世界同类型项目的成功典范。

Between 1942 and 1945 Munich was bombed by Allied air forces for more than 70 times. After the war, most citizens advocated rebuilding the old city and restoring the original face of history. In order to meet the needs of the future development of the city, Munich does not stick to the original appearance of the city in history, but constantly optimizes and reorganizes the urban structure, shapes the quality of public spaces such as squares and pedestrian streets, and improves the comfort and use function of the building. For example, while rebuilding the City Hall area, the buildings on the east and west sides were deliberately moved back by 4 to 5 meters to strengthen the artistic expression of the City Hall and the Marienplatz. The government extensively restored the street system, the façade shapes and details of major historic buildings, and the restoration of the palace continued until the 1970s. Through a series of urban construction initiatives that combined economy and people's livelihood, Munich was able to quickly regain its prosperity, revitalize the self-confidence of its citizens, and successfully won the right to host the 1972 Olympic Games. Taking the Olympic Games as an opportunity, Munich began to build a large-scale subway and gradually open up the entire area within the old city walls as a pedestrian zone. The former has further promoted the modernization level of the city and the vitality of the economy, and the pedestrian area has become a successful case of the same type of projects in the world.

自 20 世纪 60 年代已重现繁荣的慕尼黑

让大学繁荣开放共享
Letting universities open and shared

　　慕尼黑的历史城区多是由沿道路周边布置的建筑围合的街坊组成。慕尼黑大学和慕尼黑工业大学这两所世界百强大学，就是街坊里萌芽和成长起来的没有院墙的"马路大学"。和其他德国公立大学一样，这两所大学的阶梯教室是向任何对知识感兴趣的人开放的。慕尼黑大学艺术学院门厅常举办音乐会、朗诵会，吸引市民、学生和游客的参加和互动。慕尼黑大学老校区的图书馆也是一所对城市敞开大门的公共图书馆，笔者 2010 年再访慕尼黑时，在这个图书馆复印文献仍和 2002 年时一样不需要任何证件。

　　两所大学的学生宿舍都远不能满足学生的需要，使得很多学生租住在老城区各处的百姓住宅，因而他们也很自然地融入当地的生活。在老城区建设的大学自有天然的文化底蕴，而把老楼用作大学的学院，让学院根据需要修缮和更新老建筑，必然能推动建筑遗产保护和城市更新的永续发展。让精力旺盛、充满浪漫情怀的大学生成为老城区的使用者、建设者、消费者和行为艺术家，让年轻人乐此不疲，游客乐在其中，毕业生安居乐业，城市乐见其成，何乐而不为呢？

Munich's historic districts is mostly made up of block-border constructions where buildings are arranged along the road. The University of Munich and the Technical University of Munich, two of the world's top 100 universities, are the "adjoining universities" that sprouted and grew up in the block-border contructions without walls. Like other German public universities, the lecture rooms at both universities are open to anyone interested in knowledge. Concerts and recitals are often held in the foyer of the University of Munich's Faculty of Arts, which attract the participation and interaction of citizens, students and tourists. The library on the old campus of the University of Munich is also a public library that is open to the city. As I revisited Munich in 2010, I still did not need to show any documents to copy architectural documents in this library as in 2002.

The student dormitories at both universities are far from meeting the needs of students, so many students rent rooms in private homes throughout the historic district, and then they naturally integrate into local life. The universities built in historic district have their distinctive cultural charm. Using old building as the university's colleges could inevitably promote the protection of architectural heritage and the sustainable development of the old city. Let energetic and romantic college students become users, builders, consumers and performance artists, so that the historic district become a place where young people enjoy, visitors have fun, graduates stay and work... So why not?

向公众开放的慕尼黑工业大学图书馆

广场上的行为艺术家

　　慕尼黑老绘画博物馆的古典艺术珍品展一直是城市的骄傲，展出 700 余幅 13 至 18 世纪的欧洲绘画，包括著名画家达·芬奇、拉斐尔、波提切利、丢勒、伦勃朗和鲁本斯等人创作的珍品。它的建设源于巴伐利亚国王路德维希一世"大众教育"的理念，他认为有必要建设一个场所让民众欣赏到王室的藏画，让艺术珍品不再远离大众。建筑由利奥·冯·克伦泽（Leo de Klenze）设计，于 1836 年动工。老绘画博物馆不仅可以让人从绘画中体验欧洲文化的发展史，博物馆建筑也是一段特殊历史的见证。1943 年到 1944 年间，博物馆遭受过 4 次轰炸，中部损毁严重。在 1952 年至 1957 年间的修复工作中，项目主持人汉斯·多尔加斯特（Hans Döllgast）根据残留的建筑废墟进行了平面的大胆调整，改变了主入口的位置并增设了敞廊和楼梯，重新组织了展览的交通流线，使得修复后的建筑在立面上依然保留被破坏的痕迹。虽然公众对这种做法仍有争议，但这恰恰体现了德国的文物保护和建筑界认为古迹修缮应"新旧共生"的共识。

The exhibition of classical art treasures at the Old Museum of Paintings in Munich has always been the pride of the city, exhibiting more than 700 European paintings from the 13th to the 18th centuries, including those created by the famous painters Leonardo da Vinci, Raphael, Botticelli, Dürer, Rembrandt and Rubens. The construction stemmed from the concept of "mass education" by King Ludwig I of Bavaria, who believed that it was necessary to build a place for the public to enjoy the royal family's paintings, so that artistic treasures were no longer far away from the public. The building was designed by Leo de Klenze and began construction in 1836. The old painting museum can let people experience the history of the development of European culture from painting, while the museum building is also a witness to a special history. Between 1943 and 1944, the museum suffered four bombings, especially its central part. During restoration work between 1952 and 1957, the project's facilitator, Hans Döllgast, made bold plans based on the remaining architectural ruins including changing the position of main entrance, adding an open gallery and staircase, and reorganizing the traffic flow of the exhibition, so that the restored building still retained traces of destruction on the façade. Although the public still disputes this practice, it reflects the consensus of the German heritage protection and architectural spheres that the restoration of monuments should be "symbiotic with the old and the new".

慕尼黑老绘画博物馆与公共绿地

慕尼黑老绘画博物馆侧翼

寓旧于新的城市体验
Providing rich urban experience

　　慕尼黑保留原有街区格局和修缮老建筑的同时，嵌入了众多的新空间和新功能，著名的"五宫廷"就是一个典型的项目。历史街区和记忆是建筑师创作的"本底"，这意味着既尊重历史的过往，也勇于走出历史的羁绊。"五宫廷"院内新建的通廊都以与其连通的历史街道冠名，韦斯卡迪庭院还以慕尼黑统帅堂旁边的韦斯卡迪街命名，并以金色鹅卵石的铺地来纪念慕尼黑人民反对纳粹暴政的特殊事件。然而，建筑师赫尔佐格和德梅隆事务所并不"修旧如旧"，因为那样"五宫廷"自然无法超越在历史上曾有过的价值。"五宫廷"着力创造的是"寓旧于新"的艺术体验，以新鲜、高雅的格调提升街区的艺术价值。用现代、时尚的设计串联起来的五个院落是年轻人特别喜爱的地方，这里有让人目不暇接的时装旗舰店、网红书店、网红鞋店、餐具店和招贴画店，不少室内设计都是艺术设计界"大咖"的作品。比如本土设计师提塔·基斯的作品绿植走廊、北欧设计师欧拉弗尔·伊利亚森设计的大型钢网球体雕塑等。这个项目的成功表明，市中心需要有足够多的场所供年轻人来体验文化、施展才华，并且成为他们创业与社交的舞台。

While retaining the original block pattern and renovating the old buildings, Munich has embedded many new spaces and new features, and the famous "Five Courts" is a typical project. Historic districts and memories are seen as the "background" of architects' creations, which means respecting the past of history and having the courage to step out of the shackles of history. The new galleries here are named after the historic streets that connect them, and the courtyard "Viscardihof" is named with Feldherrnhalle and the golden cobblestone paving commemorates the special events of the Munich people against the Nazi tyranny. However, Herzog & de Meuron did not "repair the old as the old", because the "Five Courts" naturally cannot surpass the value that has existed in history even it is the same with what it's used to be. The project strive to enhances the artistic value of the neighborhood with a fresh and elegant style. The five courtyards connected by modern and fashionable design are particularly popular places for young people, and many interior designs are the works of famous designers. For example, the green plant corridor of local designer Tita Giese, the large steel sphere designed by Nordic designer Olafur Eliasson, etc. The success of this project demonstrates that there need to be enough places for young people to experience culture, display their talents, and become a stage for them to start a business or socialize.

五宫廷的入口

五宫廷的内街

先进的科研建筑设计
Designing advanced buildings

慕尼黑马克斯·普朗克研究院的中标方案使得场地的南侧构成形态规整、用途多元的公共空间，其主体建筑风格又完全是现代的：立面以外置遮阳板防止夏季室内过热，遮阳板后上下贯通的空腔通过"烟囱效应"促进自然通风，冬季来临时空腔又是收集太阳辐射热量、供室内取暖的温室。建筑主体和侧翼的三个中庭均呈三角形平面，建筑师在此设置逐层缩小梯段宽度的步行楼梯，增加了视觉的纵深感，消除了三角形空间可能带来的局促与逼仄的空间感受。优胜方案清晰地表明了 21 世纪的慕尼黑对待历史环境的态度：新建筑要体现新时代对舒适、节能与环保的诉求，要以合适的姿态对场所做出贡献，拒绝抄袭历史和没有创意的假古董。在笔者生活的青岛，历史城区则被巨型商业大厦一再重构了环境肌理与天际线。"五宫廷"和马克斯·普朗克研究院这类兼具时代精神和人性尺度的项目，或许能在帮助重塑街区的凝聚力以及老城区百姓对现代建筑的信心方面提供一些参考。

The winning proposal of the Max Planck Institute in Munich makes the south side of the site constitute a regular and versatile public space. Its architectural style is completely modern: the façade has an external visor to prevent indoor overheating in summer, and the cavity behind the façade promotes natural ventilation through the "chimney effect". In winter, the cavity turns into a greenhouse for collecting solar radiation and heating indoors. The three atriums of the main body and the flanks of the building have triangular planes, where the architect sets up a walking staircase that reduces the width of the stairs layer by layer, increasing the sense of visual depth and eliminating the cramped spatial feeling that the triangular plane may bring. The scheme clearly demonstrates the attitude of Munich towards the historic environment in the 21st century: new architecture should reflect the era's demand for comfort, energy and environment, contribute to the place in a suitable manner, and not to copy history and fake antiques. In Qingdao, where I live, after the historic district has been repeatedly reconstructed by giant commercial buildings in texture and skyline, projects with both time spirit and human scale like the Five Courts and the Max Planck Institute, may help recover the cohesion of the neighborhood and the confidence of the people towards modern architecture.

马普研究院创新所的细部与中庭

毗邻历史建筑的马普研究院创新所

马普研究院创新所项目建成之后，马斯塔歌剧厅前的广场成为一处活跃的音乐表演场所，每年的演出季还会搭建临时的小型实验性表演建筑。如何让临时建筑具备优秀的声学品质呢？"迷你歌剧空间 21"以其独特的外形和科技含量引起了广泛的关注。在建筑的外表面，多个三角锥体以"划破天际"的造型与历史环境形成鲜明的对比，成功地引起路人"要前来一探究竟"的浓厚兴趣。这些三角锥体的设计源于通过参数化设计软件将亨德里克斯（Hendrix）与莫扎特（Mozart）的两首作品"转译"为建筑形象。不仅如此，凸出的锥体有硬核的声学功能，吸收并向天空反射来自广场南侧街道的噪声，构筑广场的隔声屏障。借助建筑外表面的纹理，锥体的设计也和建筑结构构件的振动性质相互关联，以优化剧院内部的声学效果。来自奥地利的主创建筑师团队蓝天组建筑事务所认为他们的建筑如同一面镜子，镜像反映了城市的多样性、活力、张力和复杂性。

After the completion of the Max Planck Institute project, Marstallplatz infront of the opera house became an active music performance venue, and a temporary small experimental performance building is built during the annual performance season. How can a temporary building be created with excellent acoustic quality? "Mini Opera Space 21" has attracted widespread attention with its unique shape and technological content. On the outer surface of the building, a number of triangular cones contrast with the surounding historical environment with the shape of "breaking through the sky", successfully arousing the strong interest of passers-by to "come and have a look". The design of these triangular cones stems from the architectural image of two works by Hendrix and Mozart through parametric design software. Not only that, but the protruding cone has a hard-core acoustic function, absorbing and reflecting away the noise from the street on the south side of the square, creating a soundproof barrier for the square. With the help of material and textures, the triangular cones of the facade is also correlated with the vibrational properties of the building's structural components to optimize the acoustics of the interior. The architect team of Coop Himmelb(l)au believes architecture should be as a mirror that reflects the diversity, dynamism, tension and complexity of the city.

"迷你歌剧空间 21"的外观

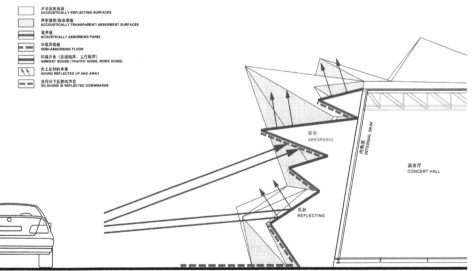

	声学反射表面 ACOUSTICALLY REFLECTING SURFACES
	声学透明/吸收表面 ACOUSTICALLY TRANSPARENT/ ABSORBENT SURFACES
	吸声板 ACOUSTICALLY ABSORBING PANEL
	半吸声地板 SEMI-ABSORBING FLOOR
	环境声音（交通噪声，工作噪声） AMBIENT SOUND (TRAFFIC NOISE, WORK NOISE)
	向上反射的声音 SOUND REFLECTED UP AND AWAY
	没有向下反射的声音 NO SOUND IS REFLECTED DOWNWARDS

"迷你歌剧空间 21"外墙的声学设计

与时俱进的住区规划
Updated residential design

在历史城区建设共享大片绿地，兼具高密度和个性化的居住空间，似乎是难以企及的事情，而特蕾莎高地（Theresienhöhe）项目做到了这一点。这个地块面积为 22 公顷，从北、西和南三面围绕巴伐利亚公园，东侧毗邻慕尼黑啤酒节举办地，这里历史上曾经是一片周边式的街坊，后成为慕尼黑博览会的所在地。随着博览功能迁出到里姆地区，这块位于市中心的土地得以重新开发。城市规划部门要求建设不少于 1400 户住宅并创造约 5000 个工作岗位。城市设计竞赛为基地未来的布局与建筑的形式提供了多种设想。希尔姆扎勒、希尔、奥纳奥纳和海茨伯格等团队都主张以周边式街坊再现市中心传统的商住模式，不同的是有的方案采用更加"纯粹"的周边式街坊，有的采用局部以独栋式或行列式建筑满足个性化的需求。中标的慕尼黑施泰德勒教授团队仅在基地的外围采用封闭的街廓，内部采用独栋式或行列式的建筑面向庭院、绿地和公园。评审委员会的最终选择表明了专家们提倡在因地制宜、保证生活品质的前提下延续城市肌理的务实态度。

Building high-density and individualized living spaces in historic districts may seem unattainable, and the Theresienhöhe project has done just that. The 22-hectare plot is surrounded by the Bavarian Park on three sides from the north, west and south, and is bordered by the Oktoberfest site on the east side. It was historically a block-border neighborhood that later became the site of the Munich Fair. With the relocation of the Fair to Riem, this centrally located land was redeveloped for no less than 1,400 dwellings and the creation of about 5,000 jobs. The urban design competition provides multiple ideas for the future layout and architectural form of the site. Teams such as Hilmer & Sattler, Hierl, Ortner & Ortner and Hertzberger all advocated recreating the traditional commercial and residential model of the city center with block-border neighborhoods, but some schemes prefer more "pure" block-border pattern, and some integrated detached or parallel patterns to meet individual needs. The winning team of Prof. Steidle in Munich used only block-border pattern on the periphery of the site, with detached or parallel buildings facing to the Bavarian Park. The final choice of the judging committee shows the pragmatic attitude of the experts to promote the continuation of the urban fabric under the premise of adapting to local conditions and ensuring the quality of life.

19 世纪末特蕾莎高地及周边城区

20 世纪末特蕾莎高地及周边城区

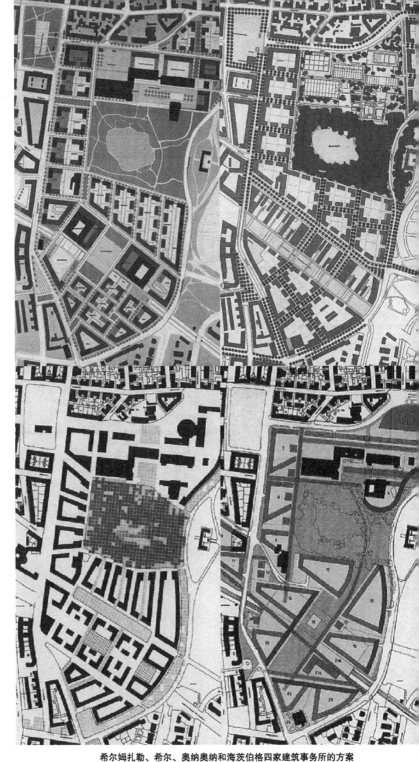

希尔姆扎勒、希尔、奥纳奥纳和海茨伯格四家建筑事务所的方案

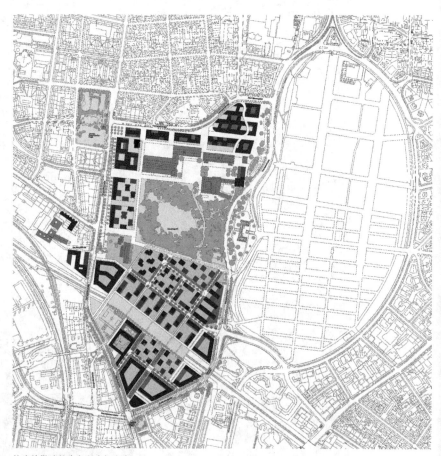

施泰德勒建筑事务所中标方案

特蕾莎高地的建筑形态

包容丰富的建筑色彩
Tolerating architectural colors

　　历史城区的新建筑应该具有什么色彩？中国有的城市专门做城市色彩研究，为各个区域的新建建筑给出特定的"色彩工具包"。德国巴伐利亚州政府的灰黑色穹顶、慕尼黑大学的砂岩色教学楼、圣母教堂的红瓦屋面和铜绿色球顶等，体现出这座城市自古以来对建筑色彩的极大包容性。对于新建筑，城市同样包容丰富的造型，甚至鼓励以色彩形成建筑和社区的个性。以前文提到的特蕾莎高地为例，在慕尼黑艺术学院施泰德勒教授的城市设计框架下，众多的居住与办公单体建筑被赋予了丰富的墙面色彩，既有暖色也不乏冷色，相邻建筑的用色"和而不同"，多样的色彩与建筑的形体共同构成了一幅具有高度审美价值的图景。在同一栋建筑上，建筑的色彩也常常混搭。比如下页图，左侧建筑白色的窗套、外遮阳卷帘与冷红色的建筑墙面形成了有趣的对比，右侧建筑立面上白色与橙色并置，显得格外活泼灵动。

What should the new buildings in the historic district look like? Some cities in China specialize in urban color research, giving specific "color toolkits" for new buildings in various areas. The grey-black dome of the German Bavaria State Government, the sandstone teaching building of the University of Munich, the red tile roof and patina dome of the Frauenkirche, etc., reflect the city's great tolerance for architectural colors. For new buildings, Munich is equally tolerant of rich forms, even encouraging the use of color to form the personality of the building and the community. In the previously mentioned Theresienhöhe, for example, under the framework of the urban design of Prof. Steidle of the Munich Academy of Arts, many residential and office buildings have been endowed with rich colors, both warm and cool, and the colors of adjacent buildings are different but harmonious. The building shapes and the diverse colors together constitute a picture with high aesthetic value. On the same building, the colors are often mixed and matched. For example, in the picture of the next page, the white window frame and outer sunshade roller of the left building form an interesting contrast with the cold red outer wall, and white and orange juxtaposition on the right building façade looks particularly lively.

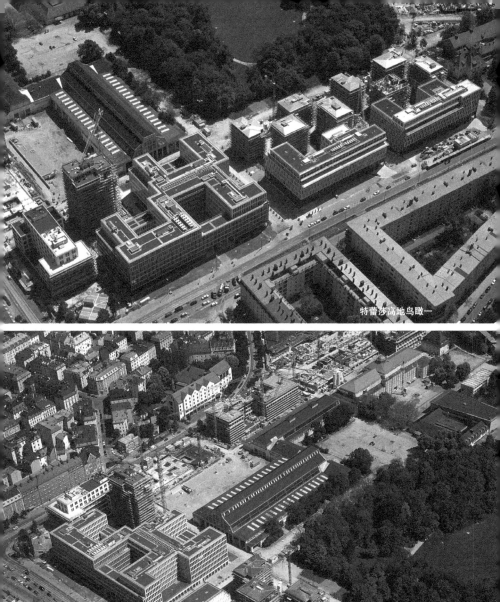

特蕾莎高地鸟瞰一

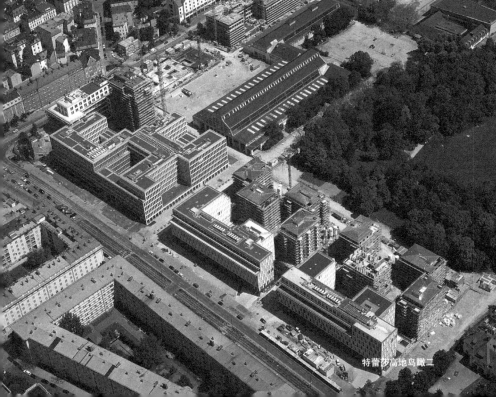

特蕾莎高地鸟瞰二

成熟社区加建养老院
Adding elderly care in mature communities

　　和我国的大城市一样，慕尼黑老龄人口占比很大，对颐养产业有巨大的需求。在历史城区见缝插针地建有电梯、集中提供照料和看护服务的养老公寓，是慕尼黑的成功经验。慕尼黑克里格建筑事务所设计的这个项目，基地位于城市主干道因斯布鲁克环路边，原有五座东西朝向的行列式住宅。这些住宅和南部住宅区的环境都受到交通噪声的干扰。政府将五栋行列式住宅的北侧串联起来，加建平行于街道的养老公寓，而加建部分对原有住宅的采光几乎不产生遮挡。建筑师以沿快速路一侧的玻璃廊组织交通，通过玻璃的隔声性能减少交通噪声对房间的干扰，而养老公寓又成为其南侧整个居住区的隔声屏障。公寓内数个房间共用一个公共客厅，充足的阳光保证建筑的热舒适性，并营造良好的室内环境氛围。养老公寓南立面的外窗上下对齐，遮阳板位置稍有错动，给整齐划一的立面造型增加了一些趣味性。到了夜晚，北立面玻璃透出的光亮使这座公寓成为富有温情和散发魅力的城市地标，为慕尼黑增添了一道美丽的风景。

Like big cities in China, Munich has a large proportion of the elderly population, and there is a huge demand for the elderly care industry. Munich began adding nursing homes with elevators and centralized care services in the old town. Architect Krieger in Munich designed the project, which is located on the side of the Innsbruck Ring Road, and originally had five east-west facing row houses. Both these houses and their southern neighbors were disturbed by traffic noise. The government connected the north side of the five row houses to build pension apartments parallel to the street, and the additional part of the original house was barely obscured. The architect organized interior flow with a glass corridor along the ring road, reducing the interference of traffic noise on the room through the sound insulation performance of the glass. The added nursing home also becomes a sound insulation barrier for the residential area to its south. Several rooms share a living room, and plenty of sunlight ensures thermal comfort in the building and creates a good flair of indoor environment. The exterior windows of the south façade are aligned up and down, and the position of the visor is slightly staggered, adding some interest to the uniform façade shape. At night, the light from the glass of the north façade makes the nursing home a warm and charming landmark.

因斯布鲁克环路加建养老院前的谷歌图片

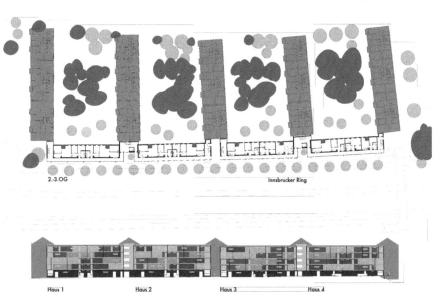

2-3.OG

Innsbrucker Ring

Haus 1   Haus 2   Haus 3   Haus 4

加建的北侧养老院部分与既有建筑形成合院

德国的养老公寓里，老人们有足够大的空间做感兴趣的事情，亲友探访时可以以很低的价格租住公寓的客房，和老人共享天伦之乐。慕尼黑在历史城区内部和周边建设的养老公寓既能够方便地利用医护资源，又便于老人和其他老龄亲友共同居住、抱团养老。在我国内，老年公寓选址在老城区的好处同样很多：首先，相对于选址在新城区或郊区，老年人对身边的环境没有陌生感，距离朋友和子女的住所往往比较近，有利于在心理上获取安全感；其次，市中心有更多和更好的医院，可以更及时地为老人提供医疗救助；再次，老城区丰富的人文环境和文化设施也便于老人们结伴游览、自得其乐，消除独处带来的孤独感。在笔者工作的青岛，历史城区"大鲍岛"有着数量众多的外廊式里院建筑，能不能让一些对老里院有情怀的老年人重新生活在大鲍岛的养老公寓，享受老街的生活、周到的关怀和家人来访的便利？这对青岛颐养产业的发展和历史城区的更新都是新的机遇，值得深入研究。

In nursing homes in Germany, the elderly have normally space to do what they are interested in. When relatives and friends visit, they can rent the rooms of the apartment at a very low price and share the joy of the elderly. The nursing homes built in and around the old city in Munich can not only easily use medical resources, but also facilitate the elderly relatives and friends to live together. In China, there are also many advantages of locating nursing homes in the historic districts: firstly, compared with the location of suburbs, the elderly have no strange feeling to the surrounding environment, and they are often closer to the residences of friends and children, which is conducive to obtaining a sense of psychological security; secondly, there are more and better hospitals that can provide medical assistance to the elderly in a more timely manner; thirdly, the cultural supplies in historic districts also facilitate the elderly to visit and enjoy themselves, eliminating the loneliness brought about by solitude. In Qingdao, where the author works, the historic area of "Dabaodao" has a large number of traditional courtyard buildings which are called "*Liyuan*". Could some of them be renovated as nursing homes? This would be a new opportunity for the development of Qingdao's elderly care industry and the regeneration of the historic district.

废弃用地变示范小区
Revitalizing abandoned area

　　"安得广厦千万间，大庇天下寒士俱欢颜"是古代诗人的美好愿景，也是当今高房价城市百姓的真切期待。慕尼黑开发的阿克曼伯根(Ackermannbogen)居住区项目不仅提供了从单身公寓到大家庭住宅的"经济适用房"，还有部分供特殊人士免费居住的公寓。项目位于奥林匹克公园南侧的地段，原来建有北约驻扎慕尼黑的军事基地，内有校场、办公建筑、士官宿舍和坦克库房等设施。冷战结束后，政府计划在这里建造一个提供654个住房单元的大型居住区。已不再是封闭的场地，如何与周边的街坊融合一起？规划和建筑如何延续场地的某种历史肌理和记忆？小区又怎样能实现良好的居住质量和步行体验？城市设计的竞标方案都对原有建筑或规划布局有不同程度的保留和利用，其中不乏引人注目的规划形态，如放射状、螺旋状、散点状，等等。从中我们也可以看出慕尼黑的新建小区并没有像中国北方城市那样严格要求新建住宅有南向的房间。由于纬度较高，慕尼黑东西朝向的住宅里主要房间都能享受日光的照耀，而不像中国南北向住宅那样有得不到阳光直射的北房间。

The Ackermannbogen residential project developed in Munich offers not only affordable housing ranging from single apartments to large family homes, but also some apartments for special people to live free of charge. Located to the south of the Olympic Park, the site was originally a NATO military base in Munich, with facilities such as a training ground, office buildings, dormitaries and a tank depot. The military standoff ceased to exist after the end of the Cold War, and the government planned to build a large residential area here that would provide 654 housing units. No longer a closed and forbidden area, how to integrate with the surrounding neighborhoods? How does planning and architecture perpetuate a certain historical fabric and memory of the site? And how can a new residential area have nice living quality and walking experience? All proposals in urban design competition have different degrees of retention and utilization of the original building or planning layout, including many eye-catching planning forms, such as radial, spiral, scattered, and so on. We can also see that the new residential areas in Munich do not necessarily require south-facing rooms as the northern Chinese cities. Due to the high latitude, the main rooms in Munich's east-west facing houses can enjoy the sunlight, unlike the north rooms of wide-spreading south-north roll houses in China, which almost do not have direct sunlight.

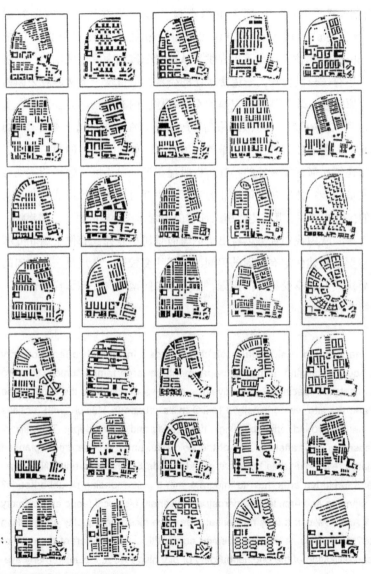

阿克曼伯根居住区部分竞标方案

阿克曼伯根项目中标方案脱颖而出，体现了德国规划界在"新建住区应尊重城市既有肌理"这一观念上的共识。北侧组团形成周边式格局，有两个原因。一是和其相邻的东侧既有街坊构成了连续的街廓，在组团边界形式上与既有环境相呼应；二是再现了场地上原有军械库房的朝向。这两个措施使得"新"与"旧"的造型元素形成了特定的相互关联。北侧组团内部的中心位置为一块长方形公共绿地，绿地两侧集中布置南北朝向的住宅。西南、东南组团延续了场地原有的东西向布局，将部分既有建筑改造为新的用途。北组团和东南、西南组团之间围合成三角形绿地，西南组团和东南组团间围合成梯形绿地，东南组团更是在原址保留了原军事基地一处长方形的广场形式。西北角的扇形绿地既再现了场所的记忆，还与基地西北侧一直延伸到奥林匹克公园的大片绿地融为一体。这些绿地和公共空间共同构成了居住区良好的景观生态和步行体验。

The winning proposal for the Ackermannbogen project reflects the consensus of the German planning community that "new settlements should respect the existing fabric of the city". The north group has a block-border pattern for two reasons. One is to form a continuous street contour with the existing neighborhoods to the east, and the other is to reproduce the orientation of the original armory on the site. These two measures result in a specific correlation between the "new" and "old" forms. The center of the north group is a rectangular green space, which is flanked by north-south facing residences. The southwest and southeast groups continue the original east-west layout of the site, and turns some existing buildings to new uses. Three groups surround a triangular green space, and the southwest group and the southeast group form a trapezoidal green space. The southeast group even retains the site and form of a rectangular square from the original military base. The fan-shaped green space in the northwest corner both recalls the memory of the place and merges with the large green space that stretches to the Olympic Park to the northwest of the plot. Together, these green spaces and public spaces make up a good landscape scenery and walking experience in the Ackermannbogen area.

阿克曼伯根居住区基地

阿克曼伯根居住区规划中标方案

　　慕尼黑马普研究院创新所的基地曾是巴伐利亚王室园林的一部分，毗邻北侧的州政府和南侧新文艺复兴风格、原为宫廷骑术学校的马斯塔歌剧厅。在基地与周边的城市肌理较二战前有了巨大改变的情况下，如何在满足建设科研建筑的同时营造出有活力的公共场所？城市设计竞赛方案中包含马普研究院创新所南侧与歌剧厅围合梯形的广场，马普研究院创新所偏居一隅而让出大片带状公共空间，以 3/4 的圆柱形体成为歌剧厅与州政府建筑之间的过渡等。中标方案的马普研究院创新所主楼北侧与州政府的南北轴线垂直，与这座历史建筑共同塑造出政府广场的仪式感；南侧则以凹字形的外轮廓和歌剧厅一同围合成长方形的公共空间序列。马斯塔广场以"动"的姿态存在，提供了举办室外表演、搭建临时剧场等多种活动的可能。与其相连的马普研究院创新所南广场则以"静"的姿态，成为一处既有仪式感又有亲民姿态的科研建筑入口空间。

The site of the Max Planck Institute in Munich is once part of the Bavarian Royal Garden, adjacent to the State Government on the north side and the Neo-Renaissance-style Opera Hall on the south side, which was originally used as the Court Riding School. With the fabric of the surrounding changing dramatically from the original appearance before World War II, how to create a vibrant public space while meeting the needs of building a research institute? In the competition, the urban design proposals included a trapezoidal plaza enclosd by the southern side of the institute and the opera hall, the institute staying in a corner to give up a large strip of open space, and the three-quarters cylindrical shape of the institute as a transition between the opera hall and the State Government. The north side of the main building of the Max Planck Institute, which won the bid, is perpendicular to the north-south axis of the state government, and together with this historic building, it creates a sense of ceremony in government square; on the south side, the concave outline of Max Planck Instituteand the opera hall form a sequence of rectangular public spaces. Martallplatz exists in a "moving" posture, providing the flexibility to hold outdoor performances and build temporary theaters. The south square of the Max Planck Institute, which is connected to it, has a "quiet" posture, becoming an entrance space for scientific research buildings with both a sense of ceremony and a people-friendly attitude.

19 世纪时的项目基地和周边环境

马普研究院创新所入围方案一

马普研究院创新所入围方案二

马普研究院创新所入围方案三

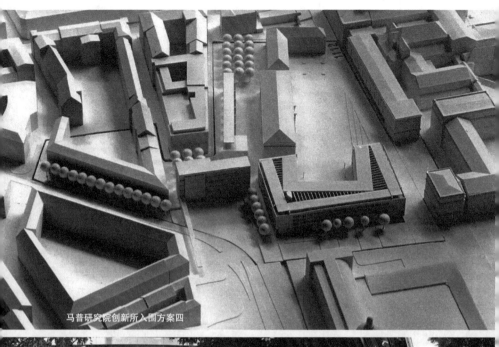
马普研究院创新所入围方案四

马普研究院创新所外景

一方面，让人们愿意在自然与人文环境中驻足、交流，是高质量城市空间的重要属性。另一方面，一个有吸引力的城市也需要在公共的节日里有盛大的活动，让亲朋好友齐聚一堂、共享欢乐。慕尼黑在休闲和交往性的环境建设上堪称世界城市的典范。从著名的马克斯米利安购物街两旁，到内城步行街、宫廷园林、遍布城市的街角公园乃至办公建筑的庭院里，都不难找到在公共座椅或台阶上阅读或交流的人们。老人在公共座椅上读报看景、中年人在公园长凳上聊天、学生和情侣在可座的台阶上交谈，成为城市中不可或缺的人文景观。每年定期举办的慕尼黑啤酒节"Oktoberfest"就是巴伐利亚州乃至整个德国都引以为傲的大型节日。轻轨和地铁把几十万市民和游客带到巴伐利亚女神像下面的特蕾莎草坪，人们在啤酒大棚里开怀畅饮，相聚在欢乐和喧闹的气氛中。各个大棚有或民族风或摇滚风的乐队，每到乐曲高潮人们就情不自禁地站到凳子、桌子上跟着节拍共同舞蹈，热烈的场面令人难忘。

The quality to let people stay and communicate within nice natural and human environment is an important attribute of good urban space. Also an attractive city needs to have places for grand events so that friends and family can gather and share the fun. Munich is a leading city in the world in terms of leisure and social environment. From the famous Maxmillian Street, to the innercity pedestrian zone, the royal garden, the pocket parks throughout the city and even the courtyards of office buildings, it is not difficult to find people reading or communicating on public chairs. Sceneries of the elderly reading newspapers on public seats, middle-aged chatting on park benches, and students talking on the seatable steps, have become human landscape in the city. The Oktoberfest, held regularly every year, is a major festival which Bavaria and even Germany is proud of. The tram and subway take hundreds and thousands of citizens and tourists to the Teresa Lawn beneath the Bavarian Goddess Statue, where people gather in beer tents in a joyful and noisy atmosphere. Each beer tent of famous brands has a national or rock music band, and every time the music climaxes, people can't help standing on stools and tables and dancing together to the beat, and the warm scene is unforgettable.

巴伐利亚州立银行的公共庭院

慕尼黑啤酒节的热闹场面

　　慕尼黑拥有数量众多、闻名遐迩的博物馆，凸显出城市作为文化与知识集聚之所的独特魅力。博物馆群落不仅让历史城区一直保持着庄重典雅的文化氛围，还吸引着源源不断的访客，带动餐饮、零售和文旅产业的发展。德意志博物馆是世界上最早的科技馆之一，也是目前世界上最大的科技博物馆，它有53个不同的博物馆区域，主题涵盖了航空航天、海洋学、物理学、天文学、化学、农业和食品技术、摄影和电影，以及航运、电信和计时技术，收藏品约12.5万件，展览流线达16公里，每年的造访者约500万。德意志博物馆的展览模式与众不同的是，它明确参展群体主要以儿童、青少年和家庭为主，系统而全面地展现着近现代科学技术发展的全景，提供了极为丰富的科普教育体验。对任何科学或工业领域感兴趣的年轻人都可以在这里找到时间跨度达几百年的珍贵展品。著名的展览项目有复原的伽利略实验室、德国发明的潜水器和飞行器、各种帆船模型等。顶层的天文馆几乎是全巴伐利亚州儿童的必去之所。

Munich has a large number of well-known museums which endow the city a unique charm as a height of culture and knowledge. The concentration of museums not only maintains a solemn and elegant cultural atmosphere in the historic district, but also attracts a steady stream of visitors, driving the development of catering, retail and cultural tourism industries. One of the world's oldest science museums and currently the largest science and technology museum in the world, Deutsches Museum has 53 different museum areas covering aerospace, oceanography, physics, astronomy, chemistry, agriculture and food technology, photography and film, as well as shipping, telecommunications and chronography technology. With a collection of about 125,000 pieces and an exhibition flow of 16 kilometers, it has about 5 million visitors per year. The exhibition model of the Deutsches Museum is different in which it clearly shows that the participating groups are mainly children, adolescents and families, systematically and comprehensively showing the panorama of the development of modern science and technology. It provides a very rich science education for young people interested in any scientific or industrial field, and they can find precious exhibits that span hundreds of years here. The planetarium on the top floor is almost a must-visit for children in all of Bavaria.

德意志博物馆

历史城区的地标项目不仅需要有高品质的建筑外观与室内空间，还需要建筑尊重历史、融入城市环境和体现相应的社会价值。慕尼黑的加斯泰格文化中心就是这一类的建筑。它位于伊萨河畔，与另一座标志性建筑德意志博物馆构成对景。作为欧洲最大的文化中心之一，它每年接待超过 200 万名访客。高负荷地使用了 30 多年后，海茵事务所赢得了加斯泰格文化中心的更新项目投标，2019 年开始承担工程的设计。建筑师在既有建筑的外侧设计了一个名为"文化舞台"的玻璃长廊，连接音乐厅、图书馆、展厅和餐厅等各个主要功能区。面向城市的透明空间将建筑内部的活力展示出来，访客与行人互为对方眼中的风景。这里容纳着音乐会、戏剧表演、艺术展览、学术会议、文化讲座和舞蹈培训等丰富多彩的活动。海茵事务所的方案不仅重塑了现代建筑的外观，也用通透的形象表达着慕尼黑开放、多元和包容的城市精神。

Landmark buildings in historic districts not only need to have high-quality architectural appearance and interior space, but also need architectural projects to respect history, integrate into the urban environment and enbody corresponding social values. The Gasteig Cultural Center in Munich is one such building. Located on the banks of the Isa River, it stands in opposition to another landmark building Deutsches Museum. As one of the largest cultural centers in Europe, it receives more than 2 million visitors a year. After more than 30 years of duty, Henn Architects won the competition for the its renewal and began to undertake the project in 2019. On the outside of the existing building, the architects designed a glass corridor called the "Cultural Stage" to connect the main functional areas such as the concert hall, library, exhibition hall and restaurant. Transparent spaces facing the city reveal the vibrancy of the building's interior, with center visitors and street pedestrians looking at each other as a sight to each other. It houses a wide range of events such as concerts, theatrical performances, art exhibitions, academic conferences, cultural lectures and dance training. The Henn proposal not only reshapes the appearance of this modern building, but also expresses Munich's open, diverse and inclusive urban spirit with a transparent gesture.

加斯泰格文化中心实施方案一

加斯泰格文化中心实施方案二

加斯泰格文化中心实施方案三

加斯泰格文化中心实施方案四

Functions   The Stage as Connecting Element   One Gasteig   Culture Stage

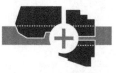

复合功能   长廊作为连接体   一个加斯泰格中心   文化的舞台

加斯泰格文化中心设计构思

加斯泰格文化中心建筑模型

黄金地段建人才公寓
Establishing talent housing in golden location

　　吸引人才是实现城市高质量发展的重要举措，而在市中心建设高技术人才公寓，也成为城市形象的载体。毗邻慕尼黑工业大学、博物馆区和伊萨河的奥斯卡·米勒公寓就是这样的一座建筑。建筑师托马斯·赫尔佐格将占建筑面积约1/3的建筑底层和顶层的空间用来作为居住者与公众学习、沟通和交流的场所，2~6层的居住空间也被压缩以增加工作室、厨房和公共休息区等交往性空间。层高6米的底层设置多功能厅、餐厅、带有图书馆的俱乐部和可举办展览的场所，在这里居住的每位国际交流学者还需要在多功能厅做公开讲座，以利于建筑的高效利用与知识的传播。

　　考虑到建筑未来也许面临扩建或改变用途，建筑师一方面在特定位置设置了易于拆除的墙体、留出扩建的接口，另一方面使管线的布置满足建筑改建为酒店时的需要。一楼多功能厅的承重柱上预留了新增楼板的插接孔，需要时可将这个高大的单层空间变为两层的空间使用。

Attracting talents is an important measure to achieve high-quality urban development, and the construction of high-tech talent apartments in the historic center can also be the carrier of the city's image. The Oscar von Miller Forum, located next to Munich University of Technology and the city's museum area, is one of such buildings. Architect Thomas Herzog used the space on the ground floor and top floor, which accounts for about 1/3 of the building area, as a place for the occupants and visitors to learn and communicate, while the private areas on the 2nd to 6th floors were reduced to increase the interpersonal space such as studios, kitchens and public rest areas. The 6-meter-high ground floor houses a multi-purpose hall, a restaurant and a club with a library. Every international exchange scholar who lives here needs to give public lectures in the multi-purpose hall to facilitate the efficient use of the building and the dissemination of knowledge.

Considering that the future of the building may face expansion or change of use, the architect on one hand left an extension interface, and on the other hand, made the layout of the pipeline to meet the needs when the building is converted into a hotel. In the multi-functional hall, plug holes on the load-bearing columns are reserved for new floor slabs, which could turn this tall single-storey space into a two-storey space when needed.

奥斯卡·米勒公寓

3. 倡导社会整合
Advocating social integration

如何有效更新那些存在较大的结构性缺陷或在社会经济方面被边缘化的街区呢？在德国，许多此类的城市更新项目的投资是市政府和州政府无力承担的。针对亟待推动的此类项目，德国 20 世纪 70 年代以来开始实行"社会城市——向街区进行投资"计划，由联邦政府、州政府和市政府每方以 1/3 的比例共同承担项目投资。三级政府通过投资，实现对项目的目标与质量的影响，共同提升项目所在区域的公共服务、空间形象与生活品质，构建适合各年龄段人群和不同收入情况的社区。近 20 年来，政府推行的"社会城市"计划更加注重发挥公共部门、市场主体与市民团体的协作。2014 年，在全德国对"社会城市"计划资助额度达到 4.5 亿欧元。本书中介绍的慕尼黑的阿克曼伯根项目、哈德恩居住区和海森贝格居住区更新项目等，就属于"社会城市"计划的一部分，同时带有慕尼黑的行政特点。在这个计划里，除了注重提升空间品质，提供更好的就业培训也是促进社会融合的规划手段。

How can neighborhoods with major structural deficiencies or socio-economic marginalization be effectively updated? Many of these urban regeneration projects in Germany are unaffordable for municipal and state governments. In response to such projects that need to be promoted urgently, Germany has implemented the "Social City" programme since the 1970s, in which the federal government, the state government and the municipal government each share the project investment in a proportion of 1/3. Through investment the three levels of government work together to improve public service, spatial image and quality of life , and build communities suitable for people of all ages and different income situations in the project area. Since the last two decades, the "Social City" programme has placed greater emphasis on collaboration between the public sector, market players and civic groups. In 2014, 450 million euros were funded in Germany for the Social City programme. Munich's Ackermannbergen project, Hadern project and Hasenbergl project, which are introduced in this book, are part of the Social City programme and also endowed with Munich's administrative features. In addition to improving the quality of space, providing better employment training is also a planning tool to promote social integration in this programme.

从慕尼黑电视塔看阿克曼伯根居住区

阿克曼伯根居住区的儿童游戏场

使衰落住区重生活力
Activating declining areas

在保留原有住户和既有建筑环境特质的前提下更新基础设施、改善住宅和公共空间，是延续历史城区文化脉络、提升经济活力和居民自豪感的有效手段。

建于 20 世纪 60 年代的慕尼黑的海森贝格（Hasenbergl）居住区位于历史城区的北侧边缘，原来的规划功能是面向低收入家庭的社会福利住宅。进入 21 世纪，海森贝格居住区的基础设施和配套服务功能已陈旧落后，老年人和移民的比例不断上升，各种社会问题凸显出来。慕尼黑政府系统地梳理了地块发展的潜力，一方面在原有规划肌理上加建配套服务功能，如地铁站、购物中心、青少年活动中心和幼儿园等，另一方面修缮和升级居住区的既有建筑与设施。通过采取小规模、渐进式的整体提升措施，既改善了生活环境的品质，创造了新的就业机会，又扭转了海森贝格居住区原本不佳的口碑和形象。今天的海森贝格不仅是德国城市更新的典范，还成为慕尼黑的文化地标和本地居民的骄傲。

Renewing infrastructure and improving residential and public spaces while preserving the characteristics of indigenous peoples and the existing built environment, is an effective means of perpetuating the historical context and enhancing economic vitality and residents' pride.

Located on the northern edge of the historic district and built in the 1960s, Munich's Hasenbergl residential area was originally planned as social housing primarily for low-income families. In the 21st century, the infrastructure and supporting services of Hasenbergl have become obsolete and backward, the proportion of elderly residents and immigrants has been rising, and social problems have become prominent. The Munich government has systematically sorted out the potential of the development of the land, then added supporting service functions to the original planning texture, such as subway stations, shopping malls, youth activity centers and kindergartens, etc., as well as to repair and upgrade existing buildings and facilities. Through the adoption of small-scale, gradual overall improvement measures, the quality of living has been improved, new jobs have been created, and the original poor reputation and image of the area have been reversed. Today, Hasenbergl is not only a successful model of urban regeneration in Germany, but also a cultural landmark and the pride of its local residents.

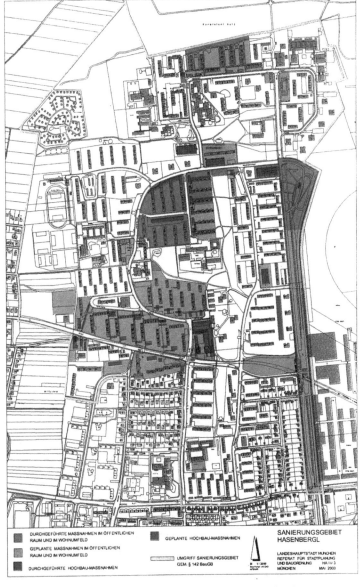

海森贝格居住区总平面图

兼顾公平的土地开发
Developing with social equality

　　慕尼黑一直在德国高房价城市榜单中名列前茅，但城市吸引年轻人和外国移民的热度仍然不减，中低收入人口对城市的满意度也比较高。究其原因，一方面是其高品质的生活环境和良好的就业市场，另一方面是其独特的土地政策，即"社会公平的土地开发"。这个土地政策的核心，是由规划项目的受益人与政府共同承担部分基础设施成本、用于特定社会性目的的住宅和就业设施的建设与运营成本，甚至该开发用地的城市设计竞赛的成本等。例如阿克曼伯根居住区项目中，规划受益人根据这项法律有义务建设市民服务中心，并参与承担幼儿园建设、道路、绿化、污染排放和环境保护的相关成本。同时，新建住宅用地的 30% 以上必须建设面向特定收入群体的住户出租或低于市场价出售的社会性住宅。这个政策从 1994 年推行至今，既有利于减轻政府建设基础设施和开展相关公共事务的经济负担，又有助于保持同一地段居民社会结构的多样性，提升中低收入群体的获得感，避免经济发达城市常出现的低收入群体在空间分布上的边缘化，在保持经济繁荣的同时保障住房供给、维持社会稳定。

Munich has consistently topped the list of Germany's highest-priced cities, but the city's popularity for attracting young people and immigrants remains unabated, and the satisfaction of low- and middle-income people with the city is relatively high. The reason is, on one hand, its high quality of life environment and good job market, and on the other hand, its land policy "socially equitable land development". At the heart of this policy is the inclusion of part of the infrastructure costs shared between the beneficiaries of the planned project and the government, including the construction and operation costs of the facilities for specific social purposes, and even the cost of urban design competitions for the development site. For example, in the Akermannsbogen project, the planning beneficiaries are obliged to build a citizen service center and participate in the costs associated with the construction of kindergartens, roads, greening, pollution emissions and environmental protection. At the same time, more than 30% of the newly built residential land must be built as social housing for a specific income group or is sold below the market price. Since its implementation beginning from 1994, this policy is not only conducive to reducing the economic burden of the government, but also helps to maintain the diversity of the social structure of residents, enhance the sense of satisfactory of middle and low income groups, avoid spatial marginalization of low-income groups that often appear in economically developed cities.

阿克曼伯根居住区和配套的幼儿园

阿克曼伯根居住区的道路

让公众参与规划决策
Decision-making with participation

公众的观点在现场展示、讨论和评估，政府部门与专家也提出意见，参与讨论。在对话中得到公众支持的措施，如果经济性与可行性经过了专业论证并符合慕尼黑市议会通过的城市发展愿景，就会在未来的几年实施。

例如，慕尼黑的管理者认为慕尼黑火车东站与周边区域的更新是城市发展的机遇，有助于解决如下的问题：使交通基础设施与经济需求相适应；提高居住与办公功能的混合程度，避免单一功能的主导；促进企业尤其是创新型企业的发展；提升公共空间的价值；以及提供市民负担得起的住房。为精准施策，政府开展了名为"行动的可能"的公众参与计划，通过所有利益相关方和政府部门之间的充分对话，确定项目需要解决的目标与问题及矛盾所在，进而为城市设计制定具体的任务。根据项目目标和所在地块的不同，参与对话的市民的年龄组成也不一样。自2016年以来，通过"行动的可能"系列对话已经为慕尼黑火车东站与周边区域更新项目的15个专题收集整理了120多项措施。

Public opinions are displayed, discussed and evaluated, and government departments and experts also provide inputs to participate in the discussion. Measures that receive public support in the dialogue will be implemented in the coming years if their economics and feasibility are professionally demonstrated and in line with the vision of urban development adopted by the Munich City Council.

For example, Munich City Government sees the regeneration of Munich East Station and the surrounding area as opportunities for urban development that help address a series of existing problems. In order to make precise measures, the government launched a public participation program called "Possibility of Action". Through dialogue, stakeholders and governmental departments identify the goals and contradictions that the project needs to address, and then develop specific tasks for urban design. Depending on the project objectives and location, the age composition of the citizens participating in the dialogue varies. Since 2016, more than 120 individual measures have been collected for 15 thematic regeneration projects of Munich East Station and the surrounding area through the "Possibilities of Action" dialogues.

4. 保护自然生态
Protecting ecological environment

慕尼黑的城市管理者认识到经济、社会、生态和文化的发展是相辅相成的，不能顾此失彼。保护城市的生态环境和公共开放空间对于保障经济活力、社会团结与文化的繁荣至关重要。对此，政府一方面严格保护森林、湿地与生物多样性，另一方面严格控制建设用地的规模，并积极开发城市中心区的工业废弃土地或因迁出机场、展览中心与军事基地而空出的用地。

　　"大型绿地序列"是一个整合慕尼黑都市区的生态发展计划，其目的是使途经历史城区的伊萨河河谷和沃尔姆河河谷绿色走廊和慕尼黑城市边缘乃至其他相邻城市之间绿化带相互连通，形成包含 14 个大型绿地在内、150 平方公里绿化面积、相互连通的大都市区绿地系统。在历史城区，人们可以近距离地来到 50 ～ 200 米宽的绿化走廊，闲暇时可以在自然景观中徒步或骑自行车来到远郊的乡野。

Munich's administration believes that economic, social, ecological and cultural developments are mutually reinforcing and cannot afford to lose sight of the other. Protecting cities' ecological environments and public open spaces is essential to ensuring economic vitality, social cohesion and cultural prosperity. In this regard, on one hand, the government strictly protects the ecological environment such as forests, wetlands and biological diversity, and on the other hand, it strictly controls the amount of construction land, and actively develops the industrial abandoned lands or the lands vacated by moving out of the airport, exhibition center and military base.

The "Sequence of large green area developments" is an ecological plan that integrates the whole Munich metropolitan area, the purpose of which is to connect the green corridors of the Isa River Valley and the Wurm River Valley, which passes through the city, and the green area between the edge of Munich and other neighboring cities, forming a metropolitan green belt system containing 14 large green spaces and 150 square kilometers of interconnected green area. In the historic district, people can get close to the green corridor 50m to 200m wide, and in their leisure time, they can easily walk or bike in the natural landscape to the countryside.

环绕慕尼黑奥体公园的公共绿地

横贯历史城区的伊萨河河谷

建设大规模生态公园
Developing large parks

    1789 年开始规划和建设的英国公园沿着伊萨河从历史城区的中心一直延伸到城市东北方向的郊野。这个慕尼黑最大的城市公园占地 3.75 平方公里，面积超过了纽约中央公园。由于公园贯穿慕尼黑历史城区，使得居民和游客可以非常方便地抵达这里。这里不仅有蜿蜒的河道和湿地、大面积的草坪和森林，还点缀着许多标志性的景观建筑。慕尼黑英国公园中著名的"中国塔"和波茨坦宫花园中的"中国亭"一样，体现了欧洲古代建筑师对东方风格的想象和向往。

    在明媚的夏日，很多慕尼黑和周边城市的市民来到英国公园散步、骑车、遛狗、晒日光浴、打排球或者安静地读书。在靠近老城墙一处古桥边的河湾"冰溪"是全世界冲浪爱好者的朝圣之地，一年四季都有勇敢者的表演。冲浪高手们往往排着队，依次踏上冲浪板滑入激流，在湍急的水流上展示自己的平衡能力和高难度的技巧，直到落入水中再优雅地游回岸上。路过这里的行人无法不让自己驻足，紧张刺激的现场表演常使人叹为观止，也让这里一直洋溢着节日般的热闹气氛。

Built since 1789, the English Park stretches along the Isa River from the city center to the countryside northeast of the city. This largest urban park covers an area of 3.75 square kilometers, even larger than New York's Central Park. As the park runs through Munich's historic district, it is very easy for residents and visitors to reach. Not only are there winding rivers and wetlands, large areas of lawns and forests, but also many iconic landscape architectures – among which there is the famous "Chinese Tower". Both the "Chinese Tower" in Munich English Park and the "Chinese Pavilion" in Potsdam Palace embody the imagination and yearning of ancient European architects for oriental styles.

On bright summer days, many citizens of Munich and the surrounding area come to the English Park for walking, biking, sunbathing, playing volleyball or quiet reading. The bend "Eisbach" near an old stone bridge is a pilgrimage place for surfers all over the world, with brave performances throughout the year. Surfers often line up to slide into the rapids, demonstrating their balance and difficult skills on the rushing currents until they fall into the water and swim back to shore gracefully. Pedestrians passing by can't help stopping, and the intense live performances are often breathtaking, and the atmosphere has always been full of festive atmosphere.

英国公园的大面积草坪和林地

伊萨河的冲浪圣地冰溪

生态修复硬化的河道
Naturalizing concrete river beds

"伊萨河计划"是恢复河道生态、优化休憩景观的最佳项目之一。伊萨河起源于阿尔卑斯山，流经奥利地蒂罗尔州和德国巴伐利亚州后注入多瑙河，是慕尼黑市区内最大的河流。出于控制雨洪和开发水电的需要，慕尼黑在20世纪中叶曾对河道截弯取直并硬化河床，使湿地的生态受到严重影响，途经慕尼黑的航运也随之萎缩。1995年起，慕尼黑着手对流经历史城区的伊萨河河道与湿地进行了大规模的重塑，具体措施包括重塑河流自然岸线、以缓坡替代滚水堰、局部以自然河床替代硬化河床和提升亲水环境等。为了防止水电站过度抽取水资源，在特别干旱的月份，政府甚至采取临时中断水力发电而避免其影响伊萨河的生态功能和休憩体验。"伊萨河计划"也使河流与城市交通网的衔接得到改善，还规划了徒步道、自行车道与机动车道结合的体验线路。慕尼黑的居民常常在夏日闲暇时来到河畔的卵石滩，在古树、绿草与流水间享受繁华都市中特有的一份悠闲。通过科学、有效的生态恢复，伊萨河的防洪与生态调节能力都得到了提升，它又重新成为自然多变的都市景观。

The "Isa Plan" is one of the best projects to restore the ecology of the river and optimize the leisure landscape. The Isa River originates in the Alps and flows through Tyrol and Bavaria before flowing into the Donau. In the middle of the 20th century, Munich straightened and hardened the riverbed due to the need to control stormwater and develop hydropower, which seriously affected the ecology of the wetlands and shrank the shipping route through Munich. Since 1995, Munich has embarked on a large-scale remodeling of the course and wetlands of the Isa River, including reshaping the natural shoreline of the river, replacing the rolling weir with a gentle slope, replacing the hardened riverbed with a local natural riverbed, and improving the hydrophilic environment. In order to prevent the over-extraction of water from hydropower plants, during the particularly dry months, the government even resorts to temporary interruptions to avoid affecting the ecological function and recreation experience of the river. The "Isa Plan" has also improved the connection between the river and the urban transportation network, and has also planned experience routes that combine hiking, cycling and motorized roads. Residents of Munich often come to the pebble beaches on the riverside in their summer time, enjoying a leisurely atmosphere unique to the bustling city. Through effective ecological restoration, the flood control and ecological regulation capabilities of the Isa River have been improved, and it has once again become an urban landscape with natural scenery and biological diversity.

重新恢复生态功能的伊萨河岸线

周末在卵石滩上休闲的人们

为骑行人士提供便利
Giving bikers convenience

　　自行车由于不污染环境、占有空间小和可锻炼身体的特性，是慕尼黑鼓励发展的出行工具。在慕尼黑历史城区，没有什么交通工具比自行车能让人更轻松地了解城市，甚至政要与明星也会骑着自行车在周末与家人穿行在市中心的英国公园里。对于轻轨与地铁和公交站点之间的短途路线，共享单车更是最便捷的交通工具。为了发展自行车交通的基础设施，整个慕尼黑市规划有超过200公里的自行车专用道，设有详细的路线标示牌，还有大量且不断增加的自行车停放点。一些重要的公共设施或者是目的地，如火车站和大学前面，总是有大量的自行车停车位，这使得自行车在游客拍摄的慕尼黑风光片中常常占有一席之地。来慕尼黑度周末的欧洲人也常常开着车顶架有自行车的私家车来到郊区停车场，再骑车出发前往市区。当然，在慕尼黑骑自行车必须有前后灯，晚上不开车灯则面临罚款：因为车灯不只给自己探路，更重要的是提示过往车辆，保证骑行安全。自行车已经是慕尼黑引以为豪的文化标识，慕尼黑历史城区也已经成为德国最适合骑车出行的地方之一。

Due to its non-polluting environment, small footprint and exercise ability, cycling is a means of travel that Munich encourages. In Munich's historic district, there is no way to get to know the city more easily than by bicycle, and even dignitaries and celebrities ride their bicycles through the English Park in the city center with their families on weekends. For short routes between the light rail subway and bus stops, renting bicycles is the most convenient means of transportation. In order to develop the infrastructure for bicycle traffic, the entire city of Munich has more than 200 km of self-employed lanes, signage for detailed routes, and a large and increasing number of parking spots. Some important public facilities or destinations, such as train stations and universities, always have plenty of bicycle parking spaces, which makes bicycles often a place in the Munich landscape photos taken by tourists. Europeans who come to Munich for weekends often drive their private cars with bicycles on the top to the suburban parking lot and cycle out to the city. Of course, cycling in Munich must have front and rear lights, and if you don't drive with lights at night, you will face a fine. The bicycle is already a proud cultural icon in Munich, and the historic district of Munich has become one of the best places in Germany to get around by bicycle.

遍布自行车专用道和停车点的历史城区

英国公园草坪的骑行者

5. 示范绿色科技
Demonstrating energy efficiency

德国在建筑节能领域一直走在世界的前列，巴伐利亚州在可持续建筑实践方面也取得了很多值得推广的经验。这些经验包括在政府投资建设项目中应用与验证节能技术和新材料，资助科研计划和与慕尼黑、奥格斯堡等城市的高校合作建立相关专业与课程等。

　　在示范住区项目和对既有建筑的改造中，政府一直充当先行者。通过投资和组织专业设计团队与高校的加入，落实节能建筑的设计和施工质量，向市场展示建筑节能的潜力与益处。在马克斯·普朗克研究院、阿克曼伯根居住区和特雷莎高地等示范项目中，政府着力打造在高效的通风、采光、遮阳和保温蓄热性能、利用可再生能源等方面的样板。在历史建筑和普通既有建筑修复中，政府也率先垂范，大力推广节能措施。慕尼黑奥运会游泳馆改造、慕尼黑市议会采光屋顶改造就是其中两个范例。慕尼黑工业大学特别设立了"气候与建筑设计"研究生专业，巴伐利亚州的建筑类院校都开设和州最高建设局合作的课程，培养建筑节能和楼宇优化技术人才。

Germany has always been at the forefront of the world in the field of building energy efficiency, and Bavaria has a lot of experience in sustainable building practices that are worth promoting. This experience includes the application and validation of energy-saving technologies and new materials in government-invested projects, funding of research projects and the establishment of relevant majors and courses in cooperation with universities in cities such as Munich and Augsburg.

The Government has been a pioneer in demonstration energy-saving in settlement projects and the renovation of existing buildings. By investing in and organizing professional teams and universities to join and ensuring construction quality of energy-saving buildings, it devotes itself to show the potential and benefits of building energy conservation to the market. In demonstration projects such as the Max Planck Institute, the Ackermannbogen project and Theresienhoehe project, the government focused on creating models for efficient ventilation, lighting, sunshading, insulation, thermal mass performance, and the use of renewable energy. In the renovation of historical buildings and other existing buildings, the government has also taken the lead in setting examples of energy-saving measures. Technical University of Munich has launched a special Master program "Climate Design", and Bavarian architecture faculties offer courses in cooperation with the state's construction authority to train technical personnel for energy efficiency and building optimization.

阿克曼伯根居住区太阳能示范一

阿克曼伯根居住区太阳能示范二

2009 年 1 月 1 日起，在建筑新建、出租和交易时必须出示建筑的能效证书。超过 1000 平方米的公共建筑的显著部位也必须公示能效证书，以便于房屋的使用者能够了解在能源方面可能支出的费用。建筑能效证书类似于家电的节能标签，以直观的方式显示建筑的年均采暖能耗、用电指标及其与政府推荐的该类型建筑节能标准的差距。证书在指标的柱状图上以不同颜色表示建筑能效的优劣：绿色代表"很好"或"好"，黄色指代"中等"，而红色表明"差"或者"很差"。同时，证书也给出在建筑的保温、通风、采光与设备等方面采取哪些措施可以提升建筑的节能水平。这种让人对建筑能效水平一目了然的证书一经推行，不仅成为房屋交易价格的重要影响因素，而且也影响到房屋的租金——节能性能不好的建筑意味着租客要承担更多的采暖和用电支出。在既有建筑改造时，建筑拟达到的能效指标是业主是否能够取得德国复兴信贷银行（KFW）贷款的条件。建筑能效证书的推行对提高业主、建筑企业和消费者的节能意识，调动全社会的力量推动建筑节能工作成效显著。

From 1st January 2009, the energy certificate of the building must be presented when it is newly built, leased and traded in Germany. Energy certificates must publicly presented in prominent parts of public buildings exceeding 1,000 square metres so that users can understand what they may be charged on energy. Building energy certificates are similar to the energy-saving labels of home appliances, visually displaying the average annual heating energy consumption of buildings, electricity consumption indicators and their gaps with the energy-saving standards of this type of building pushed by the government. The certificate indicates the strength or weakness of a building's energy efficiency in different colors on the indicator's histogram: green represents "very good" or "good", yellow means "moderate", while red indicates "poor" or "very poor". Also, the certificate gives out what measures can be taken in the insulation, ventilation, lighting and equipment of the building to improve the energy saving level of the building. Once introduced, this certificate, which provides an at-a-glance view of the energy efficiency level of the building, has become not only an important influencing factor in the transaction price, but also affects the rent of the building. When an existing building is renovated, the energy efficiency target to be achieved by the building is a condition for whether the owner can obtain a KFW loan. The implementation of energy certificates has achieved remarkable results in improving the energy awareness of owners, construction enterprises and consumers, and mobilizing the strength of the whole society to promote energy efficiency.

大学医院的公共建筑能效证书

公示建筑能效证书的医院门厅

　　阿克曼伯根居住区的北组团入选名为"太阳近热"的绿色建筑示范项目。阳光充足时，南北朝向的三栋大型建筑的屋顶通过太阳能集热器将冷水加热，热水储存在街区西侧的地下储热水池。建筑基底开挖时得到的土壤覆盖储热水池，在形成景观山丘的同时起保温隔热的作用。从储热水池中保存的能量供给大约300户居民。如果加上地源热泵提供的能源，北组团基本可以实现能源的自给自足，成为名副其实的近零能耗街区。此外，依据冬至和夏至日的阳光入射角度合理设置阳台的进深，使阳台成为房间的外遮阳设施，让建筑冬暖夏凉，也是这个街区住宅的一个设计亮点。设在窗外的卷帘不仅增强了建筑在夏季的遮阳效果，还可以在取暖季节的夜晚减少室内向室外的热辐射。依据城市的社会空间规划，阿克曼伯根居住区既有户型丰富的经济适用房和配套的幼儿园、日托中心、诊所、餐馆和超市，还有免费使用的福利住宅，以及由营房改建而来的学生公寓。

The north group of the Ackermannbogen area was selected as a model project called "Sun Near Heat". When the sun is shining, the roofs of the three large north-south facing buildings are heated by solar collectors, which are stored in underground thermal storage pools on the west side of the group. The soil obtained during the excavation of the building base covers the thermal storage pool, which plays a role in thermal insulation while forming a landscape hill. The energy stored from the thermal storage tank supplies about 300 households. If adding energy provided by the heat pump, the north group can basically achieve self-sufficiency in energy and become a nearly zero-energy block. In addition, according to the angle of sunlight incidence on the winter solstice and summer solstice, the depth of the balcony is reasonably set, so that the balcony becomes the external shading facility of the room, and the building is warm in winter and cool in summer, which is also a design highlight of the residential building in this neighborhood. Roller blinds placed outside the window not only enhance the building's shading effect in the summer, but also reduce indoor-to-outdoor thermal radiation at night during the heating season. According to the city's social and spatial planning, the residential area includes a wide range of affordable housing and a kindergarten, a daycare center, clinics, restaurants and supermarkets.

阿克曼伯根居住区的太阳能集热屋顶

小区住宅边覆盖土层的储热水池

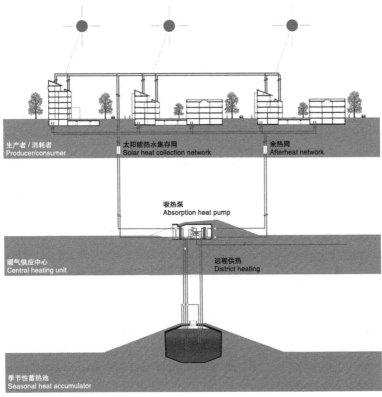

生产者 / 消耗者
Producer/consumer

太阳能热水集存网
Solar heat collection network

余热网
Afterheat network

吸热泵
Absorption heat pump

暖气供应中心
Central heating unit

远程供热
District heating

季节性蓄热池
Seasonal heat accumulator

阿克曼伯根小区太阳能储热原理

博物馆的精细化调控
Adjusting museum indoor climate

　　毗邻慕尼黑新老绘画博物馆的布兰德霍斯特博物馆，是一座由经营汉高公司的家族基金会捐赠藏品、公共财政建设的示范性绿色建筑。博物馆位于长100米、宽34米、三面临街的狭长基地上，平面呈"L"形。23种颜色、36000多根陶管垂直悬挂形成的建筑外表皮，是这座现代抽象绘画博物馆"恰如其分"的身份标志。自然光同样是塑造建筑性格的创新性元素。为保证地下展厅的自然采光，一半的展览空间的天窗开在室外平台上，天窗还可以上人，这使得室内与室外环境在光影的变化中形成互动与张力。博物馆顶层空间的自然采光更加精密与复杂，阳光通过可调控的多层过滤构造进入展厅，使室内具有合理的照度并避免出现眩光。建筑采用地源热泵作为主要热源，展厅内不设单独暖气片，均匀埋设在墙体内的暖气管道对室温进行精细的调控。通过沿墙体布置的地板通风带实现室内换风。均匀而稳定的室内气流避免了灰尘的聚积。建筑内部巨大的木质楼梯引导游客进入各层的展览空间，也起到在建筑内部衔接两个街道的作用。

The Brandhorst Museum, newly built in Munich's museum area by public investment, is a green building exhibiting the collection donated by the Henkel trust Udo and Anette Brandhorst. The museum is located on a narrow site of 100 meters long and 34 meters wide, facing three streets, with a plan of "L" shape. The façade of the building, which is suspended vertically in 23 colors and more than 36,000 ceramic pipes, is the "appropriate" identity of this museum of modern painting. Natural light is also an innovative element that shapes the character of an architecture. In order to ensure the natural lighting of the underground exhibition hall, the roofwindows of half of the exhibition space is built on the outdoor platform, which makes the indoor and outdoor environments form interaction in the change of light and shadow. The natural lighting of the top floor of the museum is more sophisticated and complex, and the sunlight enters the exhibition hall through the adjustable multi-layer filter, so that the interior has a reasonable illumination and avoids glare. The building adopts ground source heat pump as the main heat source, and there is no separate radiator in the exhibition hall, and the heating pipes evenly buried in the wall body are finely regulated to the room temperature. Indoor ventilation is achieved by floor ventilation belts arranged along the walls. Uniform and stable indoor airflow avoids the accumulation of dust. The huge wooden staircase inside the building guides visitors into the exhibition space on each floor, and also plays a role in connecting the two streets inside the building.

布兰德霍斯特博物馆沿街外观

布兰德霍斯特博物馆地下一层空间

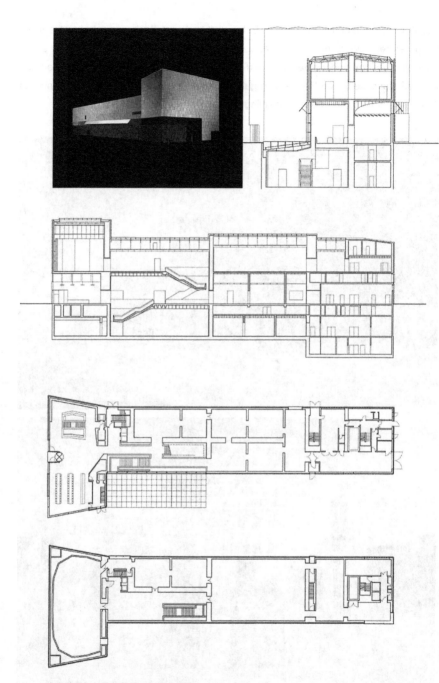

布兰德霍斯特博物馆模型和技术图纸

布兰德霍斯特博物馆屋顶采光构造

布兰德霍斯特博物馆三层空间

教堂建筑的采光设计
Design natural light in church

慕尼黑的耶稣之心教堂（Herz-Jesu-Kirche）和犹太教堂（Hauptsynagoge），是近年来建造的两个现代风格的教堂。耶稣之心教堂用了双层表皮的方式控制建筑内部的光线。经建筑外侧的玻璃幕墙引入的光线，被可转动的实木格栅构成的内墙调节，使建筑的礼拜空间沉浸在一片柔和而温暖的光线之中。位于历史城区圣·雅各布广场边上的犹太教堂是慕尼黑犹太文化中心的组成部分。粗壮的石质建筑基座旨在纪念耶路撒冷的哭墙；礼拜厅上方的玻璃顶寓意着穿越西奈沙漠露营时的帐篷，象征着犹太民族充满苦难的逃亡与迁徙历程。玻璃顶采用了三角形结构单元，让人联想起"大卫之星"的图案。慕尼黑历史城区的新建文化建筑早已不再采用历史主义的建筑语言，而是用现代的设计手法抽象地表达建筑的内涵，并与周边的环境相协调。

Munich's The Church of the Heart of Jesus and Hauptsynagoge are two modern-style churches built in recent years. The Church of the Heart of Jesus uses a double-layered skin to control the light inside the building. The light introduced by the glass curtain wall on the outside of the building is adjusted by the inner wall composed of a rotatable solid wood grille, so that the building's worship space is immersed in a soft and warm light. The synagogue on the edge of St. Jakobs Platz is part of Munich's Jewish Center. The stout stone pedestal is designed to commemorate the Wailing Wall in Jerusalem; the glass roof above the chapel symbolizes the tent as camping through the desert, symbolizing the Jewish people's journey of flight and migration, which was full of suffering. The glass roof features triangular structural elements reminiscent of the "Star of David" motif. The new cultural buildings in Munich's historic district have long ceased to adopt the architectural language of historicism, but use modern design to abstractly express the connotation of the building and harmonize with the surrounding environment.

耶稣之心教堂和广场

教堂内部

犹太教堂和广场

老王宫的棱镜形屋顶
Introducing daylights in parlarment

　　建于二战前和 20 世纪 50 年代到 70 年代之间的一万多座建筑被列入德国巴伐利亚州文物建筑法保护的范畴。巴伐利亚州的相关法律将保护的文物建筑概念定得很宽，不仅包括教堂和宫殿等重要历史和文化建筑，也包括民居、农庄、某个完整村镇及其组成部分。1973 年出台的巴伐利亚州文物建筑保护的法律要求所有者有义务对文物建筑进行保护和修缮，而且改造措施必须符合技术和艺术方面的要求。慕尼黑的巴伐利亚州议会大厦原是巴伐利亚国王马克斯米利安二世下令建造的王宫，为了让这座历史保护建筑更符合议会建筑的功能，并示范自然采光和建筑节能措施，议会大厦顶部采用了两层玻璃罩体，外层玻璃罩体只允许北侧的阳光辐射进入棱镜结构，避免南向的阳光使建筑过热，而内侧的玻璃罩体起散射光线的作用。在议会大厅开会的议员犹如置身在光线柔和的室外，并能真实地感受到外界天气与时间的变化。

More than 10,000 buildings built before World War II and between the 1950s and 1970s are classified as buildings protected by the Bavarian Heritage Building Act. The relevant laws of Bavaria define the concept of protected heritage buildings very broadly, including not only important historical and cultural buildings such as churches and palaces, but also houses, farms, a complete village or part thereof. The Bavarian Law on the Protection of Heritage Buildings in 1973 obliges owners to protect and repair the heritage buildings, and the renovation measures must comply with technical and artistic requirements. The Bavarian state parliament building in Munich was originally a royal palace ordered by King Marxmilian II of Bavaria. In order to make this historic building more in line with the function of the parliament building, and to demonstrate natural lighting and energy-saving measures, a two-layer glass cover was adopted at the top of the council. The outer glass cover only allows the sunlight from the north to enter the prism structure, avoiding the southern sunlight from overheating the building, and the inner glass cover plays the role of scattering light. Under such roof the members of the parliament are like being outside in a soft light, and they can truly feel the changes in the weather and time outside.

巴伐利亚州议会大厦及议事厅的棱镜屋顶

棱镜屋顶下以自然采光照明的半圆形议事厅

实现通风防噪的外墙
Ventilating while avoiding noise

　　建筑的外围护结构类似于人的皮肤和衣服，不仅应隔热保温，还应该有良好的呼吸作用。位于内城区（Altstadt）环路上的慕尼黑抵押银行（Hypothekenbank），为了同时解决交通噪声的干扰与建筑的自然通风问题，采用了双层玻璃箱体和普通玻璃窗并置的围护结构方案。建筑外立面的通风功能由双层玻璃箱体承担，单层玻璃窗则主要承担自然采光与引入室外景观的作用。同时，建筑也设置了机械通风系统，冬季可以将适宜温度的新风从墙体附近的地板处注入，将废气从墙体的顶棚排出。

The envelop of a building is similar to human skin and clothing, which should not only well insulate, but also play a good breathing role. The Hypothekenbank, located on Altstadt's ring road, adopts a double-glazed box and a single-glazed window juxtaposition of the envelope in order to solve the interference of traffic noise and the need of natural ventilation at the same time. The ventilation function of the building façade is undertaken by the double-glazed boxes, while the ordinary windows mainly assume the role of natural lighting and the introduction of outdoor landscape. At the same time, the building is also equipped with a mechanical ventilation system, which can inject fresh air from the floor near the wall at a suitable temperature in winter and discharge the exhaust gas from the roof.

慕尼黑抵押银行

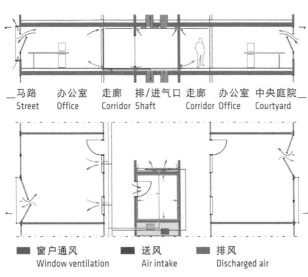

__马路	办公室	走廊	排/进气口	走廊	办公室	中央庭院__
Street	Office	Corridor	Shaft	Corridor	Office	Courtyard

■ 窗户通风　　　■ 送风　　　■ 排风
Window ventilation　　　Air intake　　　Discharged air

慕尼黑抵押银行通风方案

对于建于 20 世纪 50 年代至 70 年代的现代风格的文物建筑，政府也大力推行和示范保护性节能修缮措施。1972 年建成、1999 年列入文物建筑名录的宝马总部大楼，于 2006 年进行了全面的节能改造。建筑外墙进行了复杂的清洗，并加装了避免热桥效应与结露的内保温措施。焕然一新的宝马总部大楼既保留了原来四气缸型的特色，又完全满足当今对建筑能效和防噪声的需求。同样是 1972 年建成、已经列入文物建筑名录的慕尼黑奥运会游泳馆，通过改造屋顶的通风管理系统，使得凝聚的水汽通过天花板和顶棚之间的空间得以蒸发，既改善了建筑的整体保温性能，又延长了建筑围护结构的寿命。

For heritage buildings built in modern style from the 1950s to the 1970s, the government has also promoted energy-saving measures in their renovation. The BMW headquarter building, which was completed in 1972 and included in the list of cultural heritage in 1999, was completely renovated in 2006. The building's exterior walls were cleaned and insulation measures were added to avoid thermal bridge effects and condensation. The newly renovated BMW headquarters building retains the original four-cylinder model and fully meets today's needs for energy efficiency and noise protection. The Munich Olympic Swimming Pool, which was also completed in 1972 and has been included in the list of cultural heritage, has updated the ventilation system of the roof so that the condensed water vapor in the space between the ceiling and the roof can be evaporated, which not only improves the overall thermal insulation performance of the building, but also extends the service life of the building envelop.

慕尼黑宝马总部与植□□□□界

慕尼黑奥运会游泳馆

6. 保障建筑品质
Ensuring building quality

慕尼黑历史城区新老建筑的高质量，得益于高水平的城市规划管理、建筑设计和施工团队。"建筑与交通统筹、管理与设计结合"的体系是从巴伐利亚王国时期延续至今的传统，也被认为是巴伐利亚州的规划与建设有别于其他联邦州的地方。州一级的建设管理部门称为"巴伐利亚州最高建设局"(OBB)，隶属内政部，负责管理州内的国土空间规划、交通规划与城市建设事务，下设服务于公有土地建设的规划与建筑设计机构（类似中国国内的城乡规划设计研究院）。建筑师、交通规划师在修读了巴伐利亚州最高建设局开设的业务培训课程并考核合格后方能在建设局工作；也必须在建设局下设的规划与建筑设计机构取得工作经验后，才能从事建设管理工作。得益于规划部门的前瞻性及其与设计、研究机构间专业人才的协作，慕尼黑历史城区的城市建设与更新项目保持着很高的设计水准，对后续工作也有持续的示范价值。

The quality of old and new buildings in Munich's historic district is due to the high level of urban planning management, architectural design and construction teams. The system of "combining architecture and transport, management and design" is a tradition that has been carried down from the time of the Bavarian Kingdom to the present day, and is also regarded as the place where the construction of Bavaria is different from that of the other federal states. The construction administration at the state level, named as the "Bavarian Highest Construction Authority" (OBB), is subordinate to the Bavarian Ministry of the Interior and is responsible for the management of the urban planning, transport planning and building construction affairs in the state. It has a planning and architectural design agency serving the construction of public land (similar to the urban and rural planning and design research institutes in Chinese provinces and cities). Architects and transport planners can work at OBB only after completing the training courses and passing the examination; it is also necessary to obtain work experience in OBB's planning and architectural agency before one can work in construction administration. Thanks to the forward-looking planning authority and its collaboration with design and research professionals, the architecture and regeneration project in Munich's historic district maintains a high design standard and has continuous demonstration value for follow-up work.

慕尼黑新火车总站与南广场实施方案

慕尼黑火车总站-莱姆-帕兴中轴线改造开发

很多发展中国家的建筑工人主体是未经系统培训的农民工，这往往是造成建筑物品质和耐久性不佳的直接原因。德国则拥有以企业、学校相结合的"双元制"技工体系。由高素质的技术工人承担建筑施工，让建筑师的设计实现高水平的完成度。企业必须具有相关资质才能够成为职业学院的合作办学企业，职业学院必须将教学计划与企业中的实习实践有效地结合起来。"孩子进了技校会让父母丢脸"的意识在德国是不存在的。多数父母不干预甚至会鼓励愿意学习职业技能的子女遵从自己的兴趣。学生进入建筑类的职业技术学校不仅可以半工半读赚取生活费，而且通过职业考试并有一定年限工作经验后，同样可以进入大学攻读学位。承担建筑维护与古建修缮施工的企业里拥有过硬技术的工人收入很高，这使得行业能够吸引高素质的从业人员。我的博士生导师哈森福鲁格教授曾讲起，他在小城魏玛包豪斯街的住宅做屋面防水时，负责施工的师傅一上午就能赚两千欧元，夏天去西班牙的海岛度假不接活儿，生活水准不比教授差。

Construction workers in many developing countries are mainly migrant workers who have not been systematically trained, which is often the direct cause of poor quality and durability of buildings. Germany has a "dual system" of technician education that combines enterprises and schools, and high-quality skilled workers undertake building construction, so that architects can achieve a high level of completion of their designs. Enterprises must meet the qualifications to become a cooperative partner of vocational education, and vocational schools must effectively combine teaching programs with internship practices in enterprises. The idea that "children going to technical schools will embarrass their parents" does not exist in Germany. Most parents do not intervene and even encourage children who are willing to learn vocational skills to follow their own interests. Entering a vocational school in the construction category can not only earn living expenses with half-time work and part-time study, but also after passing the vocational examination and having a certain number of years of work experience, you can also enter the university to pursue a degree. Highly skilled workers in companies undertaking building maintenance and renovation are well paid, which enables the industry to attract high-quality practitioners. My doctoral supervisor, Prof. Hassenpflug, once said that as he was waterproofing the roof of his house on the Bauhaus Street in Weimar, the technicians in charge could earn two thousand euros in one morning, and their living standard could be better than that of the professor.

相毗邻的新老绘画博物馆优秀的施工水准

奥斯卡·米勒公寓的精致细部

　　是否应为保持历史城区的传统风貌而排斥现代的建筑与公共艺术形式？如何应对历史建筑保护与高质量生活方式之间的矛盾？21世纪伊始，慕尼黑历史城区的中环线以外开始建造高层建筑，引起了市民的广泛关注。为此，政府组织了决定高层建筑高度的公民投票，确定了新建高层建筑不得超出圣母教堂99米高度的原则。2005年起，慕尼黑城市建设部门举办了围绕"城镇风貌与新建筑"的专题研究，针对"城市的可识别性与活力""城市结构、项目选址和类型学"和"经济的可行性与生态"进行专业论证，研究的结论通过公众参与和市议会听证程序后，确定为城市建设必须遵循的法规。在德国，所有由公共资金资助的建筑研究都要向社会公布，因此人们可以在规划建设部门的网站与公共图书馆获取项目的立项决议、城市设计研究和建筑竞赛结果等信息。这保障了公众参与规划的过程中可获取完整和详实的资料支撑，专家的见解也可以被利益相关者充分知晓。经过专业而开放的城市建设与美学讨论，慕尼黑历史城区的发展遵循着这样的共识：现代建筑对城市形象做出重要的贡献，它作为经济活力、城市开放度与高质量的生活方式的载体应被合理引导，而不是受到排斥。

Should modern architecture and public art be rejected to preserve the traditional character of the historic district? How to deal with the contradiction between the preservation of historic buildings and the high quality of life? At the beginning of the 21st century, the construction of high-rise buildings in Munich attracted widespread attention from the public. The government organized a referendum to determine the height of high-rise buildings, which established the principle that new high-rise buildings must not exceed the height of the Frauenkirche cathedral. Since 2005, the Munich Urban Construction Department has held a special study on "Urban Landscape and New Architecture", which conducts special themes including "Urban Identity and Vitality", "Urban Structure, Project Site Selection and Typology" and "Economic Feasibility and Ecology", and the conclusions of the study are determined as the regulations that must be followed in urban construction after public participation and the city council hearing process. After professional and open urban construction and aesthetic discussions, the development of Munich's historic district follows the consensus that modern architecture makes an important contribution to the image of the city, and it should be reasonably guided, rather than rejected, as a carrier of economic vitality, urban openness and high-quality lifestyle.

融合众多现代建筑的历史城区风貌

已列为文物建筑的慕尼黑奥林匹克中心运动场馆

德国的建筑师事务所通常负责从提供竞标方案直到建筑验收的全过程技术服务。德国的大学培养体系和技术工人教育保证了设计与施工行业从业人员的高素质。建筑法规严格规定了各阶段的建筑设计成果的内容、深度与取费标准，保障业主与设计师双方的权益。如果建筑师的工作不能达到法律规定的标准，就会受到业主的起诉；同时，法律也对委托方的"越界"行为，如要求改动已确认的成果等，进行严格的约束。笔者在海茵建筑事务所工作期间，体验到流畅的工作流程和高效率的协作，同事乐于分享度假体验，极少有周末加班的情况。在巴伐利亚州竞赛评审专家组的组成上，遵循以建筑设计的专家、教授和相关领域专业人员占多数、业主方与政府代表占少数的方法，而且所有公开竞标项目的评审结果都书面发布在政府的纸质公告和网站上，供公众知情与监督。在城市更新的项目中，居民深度参与决策的研讨会是必不可少的，承担项目的建筑师还需要担任协调各利益相关方意见的角色。

Architects in Germany are usually responsible for the technical services of the whole building process, from the provision of the bidding proposal to the completion of the building. Germany's university system and the education of skilled workers guarantee the high quality of those working in the design and construction industry. The construction law strictly stipulates the content, depth and fee standards of architectural design results at all stages, protecting the rights and interests of both owners and designers. If the architect's work does not meet the standards prescribed by law, he will be sued by the owner; at the same time, the law also strictly restricts the "transgression" behavior of the entrusting party, such as the requirement to change the confirmed plans. During my time at Henn Architekten in Munich I experienced smooth workflows and efficient collaboration, colleagues were happy to share vacation experiences, and rarely worked overtime on weekends. The Bavarian competition judging panel is composed in a majority of experts, professors and professionals in related fields, and a minority of owners and government representatives, and the results of all public bidding projects are published in writing in the government's paper and website for public knowledge and supervision. In urban regeneration projects, workshops where residents are involved in decision-making are essential, and responsible architects also need to take on the role of coordinating the opinions of various stakeholders.

2002 年笔者在慕尼黑海茵建筑事务所竞赛部门

笔者所在团队的勃林格殷格翰制药公司项目

参考文献和图片来源
Literature and picture source

参考文献

[1] Kaeppner J., Goerl W. und Mayer C. München, Die Geschichte der Stadt[M]. Süddeutscher Verlag, 2008.

[2] Armonat T. The Isar River in Munich- Return of the Wilderness[J]. Topos, 2011, 77: 38-42.

[3] Referat für Stadtplanung und Bauordnung München. 2019. Zukunftsschau München 2040+: Szenario-Prozess und Werkstattreihe[R]. Landeshauptstadt München.

[4] Referat für Stadtplanung und Bauordnung München. 2018. Altstadtensemble München[R]. Landeshauptstadt München, 2018.

[5] Referat für Stadtplanung und Bauordnung München. 2016. Neues Wohnen in Schwabing[R]. Landeshauptstadt München, 2016.

[6] bgmr Landschaftsarchitekten GmbH. 2015. Konzeptgutachten Freiraum München 2030[R]. Landeshauptstadt München, 2015.

[7] Referat für Stadtplanung und Bauordnung München. 2015. München: Zukunft mit Perspektive[R]. Landeshauptstadt München, 2015.

[8] Referat für Stadtplanung und Bauordnung München. 2007. Innenstadtkonzept[R]. Landeshauptstadt München, 2007.

[9] Referat für Stadtplanung und Bauordnung München. 2007. Modellprojekt: Rund um den Ostbahnhof – Ramersdorf – Giesing[R]. Landeshauptstadt München, 2007.

[10] Munich Department of Urban Planning and Building Regulation. PERSPECTIVE MUNICH – Strategies, Principles, Projects[R]. Munich, 2002.

[11] Oberste Baubehörde im Bayerischen Staatsministerium des Innern. Siedlungsmodelle, Neue Wege zu preiswertem, ökologischem und sozialem Wohnungsbau in Bayern [C]. München, 2000.

[12] Serge Salat. 城市与形态：关于可持续城市化的研究 [M]. 香港：香港国际文化出版有限公司，2013.

[13] 易鑫，克劳斯·昆茨曼（Klaus Kunzmann）. 向德国城市学习：德国在空间发展中的挑战与对策 [M]. 北京：中国建筑工业出版社，2017.

[14] 罗伯特·贡萨洛，卡尔·J·赫伯曼. 建筑节能设计——从规划到施工 [M]. 北京：中国建筑工业出版社，2008.

[15] 李晋. 再认识 荷兰现代建筑对谈录 [M]. 北京：中国建筑工业出版社，2021.

[16] 刘崇. 漫谈青岛和欧洲的周边式街坊 [R]. 山东科技大学学术报告. 青岛，2022-5-20.

[17] 刘崇. 包豪斯百年与德国现代建筑的节能改造 [R]. 2019 上海国际保温防水展学术报告. 上海新博览中心，2019-12-11.

[18] 刘崇. 博物馆中的城市和城市中的博物馆 [R]. 2019 "世界博物馆日"学术报告. 青岛德国总督官邸旧址博物馆，2019-5-18.

[19] 刘崇. 青岛与慕尼黑历史街区发展的比较与思考 [R]. 2018 中德历史城区保护与可持续发展青岛论坛学术报告. 青岛八大关小礼堂，2018-12-9.

[20] 刘崇. 青岛和慕尼黑：对城市可持续发展的思考 [R]. 2017 "世界城市日"学术报告，青岛市城乡规划展示中心，2017-10-31.

[21] 巴伐利亚内政部最高建设局，慕尼黑工业大学. 2008 高能效的建筑设计与施工展览资料 [Z]. 慕尼黑，2008.

图片来源

封面上 慕尼黑鸟瞰 摄影师 Karl Schillinger

海茵事务所中标西湖大学 海茵建筑事务所

自 20 世纪 60 年代已重现繁荣的慕尼黑 摄影师 Karl Schillinger

"迷你歌剧空间 21"的声学设计 蓝天组建筑事务所

希尔姆扎勒、希尔、奥纳奥纳和海茨伯格四家建筑事务所的方案 希尔姆扎勒、希尔、奥纳奥纳和海茨伯格建筑事务所

施泰德勒建筑事务所中标方案 施泰德勒建筑事务所

特蕾莎高地建筑形态 上图施泰德勒建筑事务所 Johannes Ernst，左中图 Investa 公司，左下图摄影师 Franziska von Gagern，右下图摄影师 Franziska von Gagern

特蕾莎高地建成鸟瞰一 摄影师 Reinhard Görner

特蕾莎高地建成鸟瞰二 摄影师 Reinhard Görner

因斯布鲁克环路加建养老院前的谷歌图片 谷歌地图

加建的北侧养老院部分与既有建筑形成合院 慕尼黑 GEWOFAG 股份公司、克里格建筑事务所

阿克曼伯根居住区部分竞标方案 巴伐利亚州内政部最高建设局

阿克曼伯根居住区基地 巴伐利亚州内政部最高建设局

阿克曼伯根居住区中标方案 巴伐利亚州内政部最高建设局

20 世纪 30 年代的马斯塔广场及周边 巴伐利亚州内政部最高建设局

马普研究院创新所入围方案一 巴伐利亚州内政部最高建设局

马普研究院创新所入围方案二 巴伐利亚州内政部最高建设局

马普研究院创新所入围方案三 巴伐利亚州内政部最高建设局

马普研究院创新所实施方案 巴伐利亚州内政部最高建设局

加斯泰格文化中心实施方案一 海茵建筑事务所

加斯泰格文化中心实施方案二 海茵建筑事务所

加斯泰格文化中心实施方案三 海茵建筑事务所

加斯泰格文化中心实施方案四 海茵建筑事务所

加斯泰格文化中心构思概念 海茵建筑事务所

加斯泰格文化中心建筑模型 海茵建筑事务所

海森贝格居住区总平面图 巴伐利亚州内政部最高建设局

阿克曼伯根小区太阳能储热原理 巴伐利亚州内政部最高建设局、慕尼黑工业大学和业主

布兰德霍斯特博物馆模型和技术图纸 绍尔布鲁赫和哈顿建筑事务所和业主

犹太教堂和广场 梁雨霞

巴伐利亚州议会大厦及议事厅的棱镜屋顶 巴伐利亚州议会图片档案，摄影师 Rolf Poss

棱镜屋顶下以自然光照明的半圆形议事厅 巴伐利亚州议会图片档案，摄影师 Rolf Poss

慕尼黑抵押银行通风方案 豪斯拉登工程师事务所

慕尼黑新火车总站与南广场实施方案 奥尔韦伯建筑事务所

火车总站 - 莱姆 - 帕兴中轴线改造开发 慕尼黑市和巴伐利亚州内政部最高建设局

融合众多现代建筑的历史城区风貌 摄影师 Karl Schillinger

后记
Postscript

　　分享在慕尼黑的生活经历以及读博和任教期间的数次访学体验，是笔者一直怀有的心愿。感谢德国包豪斯大学迪特·哈森福鲁格教授指导笔者建立中西城市文化的比较视野，并推荐笔者加入慕尼黑海茵建筑事务所的设计团队。笔者在慕尼黑期间曾住在事务所竞赛部主任建筑师阿希姆·巴德先生位于施瓦本区的家里约有半年之久，得到了巴德夫妇的关爱和友谊；在青岛理工大学任教后得到特聘教授、曾任巴伐利亚州最高建设局规划负责人和慕尼黑工业大学教授的赫伯特·卡尔迈耶先生的诸多指导与帮助。对巴伐利亚州城乡可持续发展的介绍是卡尔迈耶教授在山东交流期间的重要贡献。巴德先生和卡尔迈耶教授都已仙逝，非常怀念他们。

　　感谢徐飞鹏、郝赤彪、梁雨霞等诸多师友对笔者研究与教学的帮助，感谢德国同行凡尔纳·鲍耶勒、约翰·彼得·帅克和克劳斯·昆茨曼等专家的协力支持。

　　感谢工作室研究生李含悦、钱馨仪等同学的编辑和校对工作。特别感谢中国建筑工业出版社的编辑对本书文字编审工作的辛勤付出。感谢国家自然科学基金项目 (51178228) 给予笔者团队的研究资助。经验所限，书中纰漏在所难免，恳请提出宝贵的意见。

刘崇

2022 年 6 月 29 日

It has always been my wish to share my experience in Munich during my work time and several academic visits. I would like to thank Prof. Hassenpflug, my doctoral supervisor at the Bauhaus University Weimar in Germany, for guiding me to establish a comparative vision of Chinese and Western urban culture, and for recommending me to join the design team of Henn Architects in Munich. During my work time in Munich, I lived in the downtown home of Mr. Joachim Bath, the director of the competition sector at Henn Architects, for about half a year with care and friendship of the Bath family. At Qingdao University of Technology, I deepened my knowledge of German urban construction from Guest-professor Herbert Kallmayer, who worked both as the director of the urban planning department at the Supreme Construction Bureau of Bavaria and as professor at the Technical University of Munich. The introduction on the sustainable development in Bavaria is an important contribution of Prof. Kallmayer during his exchange in Shandong. Mr. Bath and Prof. Kallmayer have passed away, and I miss them very much.

I appreciate Prof. Xu Feipeng, Prof. Hao Chibiao and Ms. Liang Yuxia and many other colleagues for their help in my work, and appreciate Prof. Werner Bäuerle, Prof. Johann-Peter Scheck and Klaus R. Kunzmann for their supports from the German side.

We are grateful to Xinyi Qian and other students for their supporting work, and especially to editors of China Construction Industry Press, for her hard work in text editing and reviewing this book. We are grateful to the National Natural Science Foundation of China (51178228) for funding the research of the author team. Due to the limitations of experience, it is inevitable to make mistakes in this book, and we kindly ask for valuable comments.

Liu Chong
June 29th, 2022